Contributions to Environmental Sciences & Innovative Business Technology

Contributions to Environmental Sciences & Innovative Business Technology (CESIBT) is an interdisciplinary series of peer-reviewed books dedicated to addressing emerging research trends relevant to the interplay between Environmental Sciences, Innovation, and Business Technology in their broadest sense. This series constitutes a comprehensive up-to-date interdisciplinary reference that develops integrated concepts for sustainability and discusses the emerging trends and practices that will define the future of these disciplines.

This series publishes the latest developments and research in the various areas of Environmental Sciences, Innovation, and Business Technology, combined with scientific quality and timeliness. It encompasses the theoretical, practical, and methodological aspects of all branches of these scientific disciplines embedded in the fields of Environmental Sciences, Innovation, and Business Technology.

The series also draws on the best research papers from EuroMid Academy of Business and Technology (EMABT) and other international conferences to foster the creation and development of sustainable solutions for local and international organizations worldwide. By including interdisciplinary contributions, this series introduces innovative tools that can best support and shape both the economical and sustainability agenda for the welfare of all countries, through better use of data, a more effective organization, and global, local, and individual work. The series can also present new case studies in real-world settings offering solid examples of recent innovations and business technology with special consideration for resolving environmental issues in different regions of the world.

The series can be beneficial to researchers, instructors, practitioners, consultants, and industrial experts, in addition to governments from around the world. Published in collaboration with EMABT, the Springer CESIBT series will bring together the latest research that addresses key challenges and issues in the domain of Environmental Sciences & Innovative Business Technology for sustainable development.

Aloysius Edward J. · K. P. Jaheer Mukthar ·
Madhu Dhruvakumar · T. K. Murugesan
Editors

Digital Transformation for Business Sustainability

Trends, Challenges and Opportunities

Springer

Editors
Aloysius Edward J.
Faculty of Commerce and Management
Kristu Jayanti College Autonomous
Bengaluru
Bengaluru, Karnataka, India

K. P. Jaheer Mukthar
Department of Economics
Kristu Jayanti College Autonomous
Bengaluru
Bengaluru, Karnataka, India

Madhu Dhruvakumar
Kristu Jayanti College Autonomous
Bengaluru, Karnataka, India

T. K. Murugesan
Kristu Jayanti College Autonomous
Bengaluru
Bengaluru, Karnataka, India

ISSN 2731-8303 ISSN 2731-8311 (electronic)
Contributions to Environmental Sciences & Innovative Business Technology
ISBN 978-981-99-7057-5 ISBN 978-981-99-7058-2 (eBook)
https://doi.org/10.1007/978-981-99-7058-2

This Springer imprint is published by the registered company Springer Nature Singapore Pte Ltd.
The registered company address is: 152 Beach Road, #21-01/04 Gateway East, Singapore 189721, Singapore

Contents

Chapter 1
One Planet One Compliance: Stakeholder Perception Analysis on the Efficacy of the IR Framework of ESG Disclosure

Musarrat Ara and B. Harani

Abstract The objective of conducting this research is to investigate the efficacy of <IR> Framework in ESG disclosure with the help of supporting literature as well as by conducting stakeholder perception on the Integrated Reporting framework through the questionnaire method. For this, an initial attempt has been made to reveal whether there is One Planet One Compliance. And then we have discussed the perception of stakeholders on the efficacy of <IR> in ESG disclosure. The study reveals that though there are diverse reporting frameworks prevailing, each has its own efficiency like CDP aligned with TCFD and CDSB possesses its own uniqueness and is highly efficient for disclosing climate-related issues. While, GRI and SASB provides standard & metrics across all the three pillars of ESG for diverse industries. Besides, <IR> is efficient in providing guidance and principles on 'what to be reported and to what extent'. Additionally, from the stakeholder perception analysis, it was concluded that <IR> is an 'umbrella' for corporate reporting and capable of replacing other reporting approaches as it is 'Extremely Effective' in disclosing the Business model, its value creation process, Risk and Opportunities, Management dialogue on emerging environmental and social risk/opportunities (e.g. climate change, HSE, Human rights, etc.). Thus, a collaboration of these organisations under one umbrella of Value Reporting Foundation and subsequently merging with IFRS Foundation is a pathway towards One Planet One compliance.

Keywords Integrated reporting framework · ESG · Stakeholder perception · Reporting sustainability

M. Ara (✉) · B. Harani
Reva University, Bengaluru, India
e-mail: Musarrat.ihtbhilai@gmail.com

A. Edward J. et al. (eds.), *Digital Transformation for Business Sustainability*, Contributions to Environmental Sciences & Innovative Business Technology,
https://doi.org/10.1007/978-981-99-7058-2_1

1.1 Introduction

1.1.1 Theoretical Background

Corporate sectors are continuously looked at as agents of revolution in an economy facing intensifying environmental in addition to social issues. Companies are incorporating these issues in their strategy and due to the consequence of internal as well as external stakeholder pressure (Haruna and Mkhize 2020; Martínez-Ferrero et al. 2016), they are going beyond the stringent compliance, adhering to environmental, social and other local regulations (Baron 2014). Thus, the companies want to inform their stakeholders about their responsible business processes, practices, strategy, commitment and the initiatives taken to manage those issues (Hoang 2018). Consequently, the corporate disclosure of ESG has emerged over the past few decades in a variety of directions.

The ESG disclosure is the reflection of the internal process of a company, its strategy, its durability, performance and its impact, emphasising non-financial metrics (Park and Oh 2022), but has financial consequences too, like: capital access, productivity and cost savings, risk management, human capital, revenue growth, etc. (UNSSE 2015). As ESG factors are continuously evolving, companies are practicing different varieties of reporting methods as per the industry and the country they belong to or have their operating segments. As a result, adopt the existing reporting practices that meet the expectations of stakeholders or follow national securities listing obligations for disclosing material information or the framework that has an international appearance (UNSSE 2015); accordingly, stakeholders are bound to refer to multiple annual reports in their decision-making process.

Therefore, in this research, we intend to discover the answer to '**One Planet One Compliance?**' for ESG discourses, by studying the multiple reporting frameworks, guidelines and supporting research literature. And so, we will be firstly discussing multiple reporting frameworks prevailing, globally; secondly, we will discuss the endeavour of the Value Reporting Foundation for one compliance and last we will be presenting the results of stakeholder perception analysis conducted on the efficacy of <IR> framework (which is assumed to be most effective and innovative framework).

1.1.2 One Planet One Compliance? ESG and Multiple Reporting Framework

While ESG disclosure is a matter of materiality and bringing transparency to the stakeholder to achieve legitimacy and stewardship (Ara et al. 2020; Haruna and Mkhize 2020), however, there still remains a disparity between what is to be reported and which information will satisfy the stakeholder needs (Singaraju et al. 2016). Thus, international organisations, e.g. GRI, IIRC, CDSB, TCFD, SASB, etc. are coming up with multiple innovative frameworks as well as guidelines for the disclosure of

ESG data. The Expert Group of UNCTAD mentioned that the stock exchange can play an influential role in promoting diverse and high-quality disclosure of voluntary sustainability disclosures (UNCTAD 2012). For instance, the stock exchanges of South Africa and Brazil created a leading example that the stock exchanges of more developed countries from China to India and even France have learned lessons and subsequently started acquainting new sustainability reporting guidelines (UNCTAD 2012).

Consequently, in 2009 SSE (at present its UN-Sustainable Stock Exchange initiative) was propelled in collaboration with UNCTAD, UNGC, UNEP and PRI to volunteer foremost stock exchanges with the public commitment to stimulate corporate reporting and ultimately on ESG issues (UNCTAD 2012). As per the database of UNSSE, sixty-seven stock exchanges have ESG reporting instruments (level of reference for each guideline/framework has been shown in chart-1 below), which has been discussed briefly in this section.

Global Reporting Initiative (GRI): The Topper of ESG Reporting: Following public uproar over the environmental destruction caused by Exxon Valdez: the oil spill disaster, GRI was established in collaboration with Tellus Institute, UNEP and the CERES (GRI 2023) with the first aim of crafting a safeguard mechanism and to make companies abide by responsible environmental conduct and follow responsible business principles, later these principles extended to cover governance and social issues for sustainable economic development to attain UN-SDGs. From the first version, G1 (2000), GRI has evolved in multiple directions and is continuously improving with subsequent releases of G2 (2002), G3 (2006), G4 (2013) and transitioned to The Universal Standard (2021) which includes Tax indicators and comprehensive waste metrics as well as rolled out Sector Standards (GRI 2023). Consequently, GRI became a highly reporting standard, i.e. 4 out of 5 companies adopt this standard for sustainability reporting (KPMG 2022) and 96% of stock exchanges refer to GRI (UNSSE 2023).

SASB: The Investors' choice for financial ESG material information: Developed in 2011, SASB standards are more investor-focused, i.e. it provides guidance for disclosing financial-material ESG information to the investors. The coverage of the SASB standard is broad and industry-based for seventy-seven industries that identify the material sustainability impact on the financial performance of each industry (SASB 2023). As per (UNSSE 2023), 79% of stock exchange referred to SASB, and is popularly reporting standard in the US, Brazil and Canada (KPMG 2022).

The <IR> Framework: for Value Creation through Integrated Thinking: Since the establishment of IIRC, the adoption of <IR> Framework and the presentation of integrated reports have gained considerable fame around the world (IIRC 2019a, 2020). The primary objective of the framework is to promote stewardship, accountability, interconnection and interdependencies for various capitals (Natural, Social&Relationship, Manufacturing, Human, Intellectual and the financial capitals) (IIRC 2021) and how the organisation implement integrated thinking using these six capitals to create value over time, which makes this framework different from other sustainability reporting standard.

CDP: A Carbon Disclosure Project for environmental impact: CDP with an outreach of 18,700+ companies in 1100+ cities runs the global environmental disclosure system to determine and manage companies' risks and opportunities related to climate change, deforestation and water security on the request of city stakeholders, investors and purchasers (CDP 2023a). It consists of a comprehensive dataset on three major environmental issues, viz. climate change, water and forest, that enable companies to uncover environmental risks and opportunities, boost competitive advantage, improve reputation and stand in tracking progress towards a sustainable economy for people and the planet (CDP 2023b). Its participants are not only companies, even sub-national governments of 96 states utilises this global platform to showcase their transition to a resilient and low-carbon economy. Consequently, organisation such as Climate Group, UNFCCC's -Climate Action Portal use this platform and also efficient for major projects e.g. RegionsAdapt, United Nations' Race to Resilience; Race to Zero and EU Climate Pact (CDP 2023c).

TCFD: Climate-related financial disclosure for effective capital allocation: Due to rapidly increasing global warming and subsequently the Paris Agreement, nearly 200 governments decided to discourse on climate change by holding the average global temperature vigorously below 2 °C and pursuing efforts to control the increase of 1.5 °C (UNFCCC 2015). Thus, in 2017, the Financial Stability Board released its final recommendation on climate-related financial disclosure, i.e. TCFD recommendations. These recommendations are designed to obtain forward-looking information around four thematic areas: governance, strategies, risk management, and most importantly, metrics and targets so as to support investment, insurance underwriting decisions, lending and ultimately improving understanding of climate-related physical as well as transitional risk within 2 °C and/or 1.5 °C scenario analysis (TCFD 2022). Due to this efficiency, 63% of stock exchanges recommend this framework for Climate-related disclosures (UNSSE 2023).

CDSB: For Natural, Human and Social Capital disclosures: The framework consists of guiding principles with other reporting requirements in the context of environmental and social disclosures such as E-S-related risk and opportunities, outlook, performance, results and impacts that are required to be presented in mainstream reports, i.e. annual reports. This framework is hosted by CDP and also encompasses the definitions of IIRC's Natural capital, 'provider of financial capitals' and similar to <IR> framework to some extent (CDSB 2023). Hence, its popularity is less, and only 36% stock exchange recommends this for ESG disclosures (UNSSE 2023).

Merger of the Value Reporting Foundation with IFRS: A pathway towards *One Planet One Compliance*: From the above discussion, it was discovered that there are inevitably diverse voluntary frameworks, guidelines and standards among which finding comparability and consistency becomes a major challenge. Hence, in June 2021, the two entities IIRC and SASB laid the foundation of The Value Reporting Foundation (henceforth, The VRF) with the intention to address those challenges and to simplify the system of disclosing ESG information. The VRF offered an ample suite of resources e.g. the <IR> Framework, Integrated Thinking Principles, and the SASB Standards to aid business and the investors to understand how the enterprise value is created, preserved and/or eroded overtime (SASB 2023).

However, due to urgent action on climate change and other sustainability issues, the IFRS Foundation publicises three significant developments in COP26: First, reformation of ISSB to construct a comprehensive universal baseline to leverage transparency, high-quality, reliability and comparable reporting mechanism; second, consolidation IFRS foundation with CDSB (hosted by CDP) and The VRF, with a commitment for investor-centric ESG disclosures; third, publication of Prototype developed by TRWG (Technical Readiness Working Group is the joint work of the IASB, CDSB, TCFD, IOSCO, WEF, VRF and the Expert Group of securities regulators (IFRS 2021); for climate-related (CDSB 2023; TCFD 2022) and general disclosure requirements (SASB 2023b).

Inevitably, the new ISSB and the reformed IFRS foundation shall be efficient in reducing the over-burdening of corporate reporting and yield high consistency and better comparability in ESG disclosures in the future and in the long run.

1.1.3 Review of Literature and Hypothesis Development

<IR> Framework Encompassing ESG Disclosure. The transition in corporate reporting is believed to be due to the theory of stakeholder (Hoang 2018); diffusion of innovation (Ara et al. 2020), regulatory pressure, theories of neo-institutionalism (Gelmini and Vola 2021) and ultimately to achieve legitimacy (Haruna and Mkhize 2020) and improve market reputation (Singaraju et al. 2016).

Still, the investment community, i.e. investors are increasingly looking for more transparency and accountability (Setyawan and Kamilla 2015) towards the sustainable development of the whole economy (Cosmulese et al. 2019; IFRS 2021; Sridharan 2018). Obviously, to meet this demand, corporations are looking for an efficient platform and a framework that allows them to provide holistic revelation of their financial and non-financial information. As a result, <IR> has gained the attention of 76% of the stock exchange (UNSSE 2023) with broad coverage of 70+ nations and 2500+ companies of different industries (IIRC 2021). It is considered as an ideal platform for ESG disclosures (Corvino et al. 2020); even 70% (N = 1434) of TCFD recommendations are reported in Integrated Reports (TCFD 2022).

In recent research, the evidence reveals that corporate governance (Hoang 2018; Ofoegbu et al. 2018) is the factor that influences the level of environmental and social disclosures (Bernardi and Stark 2016; Umoren et al. 2015). A significant upsurge was seen in South African companies' corporate governance disclosure after mandating <IR> framework (Ahmed Haji and Anifowose 2017) and the companies that have environmental committee provides more environmental disclosures in their integrated reports (Ofoegbu et al. 2018). Additionally, in another study, it was confirmed that <IR> could be an efficient tool in disclosing human rights issues as well as health and safety by improving the overall corporate governance structure (Corvino et al. 2020).

Although, the ultimate audience of any reporting framework is investors (Mähönen 2020), i.e. the provider of financial capital only (IIRC 2013; Park and

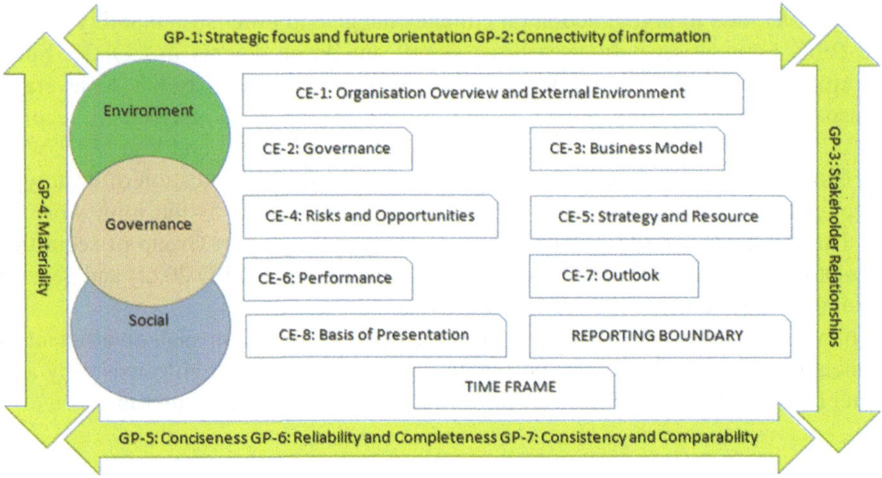

Fig. 1.1 IR > Framework Encompassing ESG Disclosure (Where GP = Guiding Principle; CE = Content Element. *Source* Own Compilation

Oh 2022). In South Africa, there was a significant association of analyst forecasts with environmental and governance disclosure, viz. post-mandating <IR> regime (Bernardi and Stark 2016). Additionally, the study also provides evidence that the integrated reports are effective in providing useful ESG information that links to financial performance which is useful for the capital market. These findings coincide with the findings of Bartha et al. (2017), Lee and Yeo (2015), Zhou (2017). Furthermore, in the study of integrated reports of Nigerian companies, it was found that the <IR> framework is efficient in governance (81%) and social (66%) disclosure (Umoren et al. 2015). Whereas, there is a positive effect of corporate governance indicators, viz. board gender, board size and CSR committee on the environmental disclosures presented in integrated reports (Raimo et al. 2021).

With the support of the above literature, we propose the following model for <IR> encompassing ESG disclosure (Fig. 1.1) and subsequently state the hypothesis:

H_0: <IR> is not an umbrella for corporate reporting and could not replace other reporting frameworks in India.

H_1: <IR> is an umbrella for corporate reporting and could replace other reporting frameworks in India.

1.1.4 Objective of the Study

The primary objective of this research is to reveal the efficiency of <IR> framework in disclosing ESG parameters. Therefore, the following research questions have been articulated:

RQ 1: What are the limitations of today's corporate reporting system?
RQ 2: How effective is the <IR> Framework in ESG disclosure?
RQ 3: Is <IR> framework efficient to replace Sustainability Report, Annual Report and BRR?

1.1.5 Scope of the Study

The study brings awareness about the efficiency of GRI, SASB, CDSB, CDP, TCFD, and more centric to <IR> Framework with empirical evidence that will help regulatory bodies (especially SEBI) to think on mandating the framework for top listed (at least first 1000) companies in India. Additionally, the study will also encourage organisations to adopt <IR> for ESG reporting. Last but not limited, the study would be a major contribution to Indian literature in the research of <IR> framework providing avenues for researchers to conduct further research on this framework.

1.2 Research Methodology

1.2.1 Research Design

With the help of a five-point Likert scaling method, the articulated objectives have been furnished by qualitative analysis based on primary data collected from various stakeholders.

1.2.2 Purposive Sampling and Data Collection Methods

The theoretical background of this research is supported by secondary data, whereas all the research questions were studied with the help of primary data only.

A systematically structured questionnaire with five-point Likert scaling consisting of thirty-two questions was administered to 500+ responders in the selected sample size through mail, out of which 202 responses had been received in 2020–21 and was considered as the final sample for this study. The questionnaire was administered to corporate people working within Integrated Reporting, CSR, Sustainability Reporting, sustainability department, accounts and finance department, ESG Consultant/Specialist and Sustainability Analyst.

1.2.3 Source of Data

In this study, all the research questions defined under objective are studied through primary data, i.e. stakeholder perceptions. However, the theoretical background of this research is supported by secondary data and previous literature.

1.2.4 Data Analysis Plan

Firstly, the Descriptive Method has been used to analyse the response rate for RQ 1, RQ 2 and RQ 3. Additionally, Cross Tabulation with a Chi-Square Test and Cramer's V has been executed to test the hypothesis and understand the strength of the association between two variables (defined below).

IR_Umbrella_Framework_Sentiments = Q. To what extent do you agree that integrated reporting is an 'umbrella' for corporate reporting, providing the context and linkage for other forms of reporting in India?

IR_Replace_Sentiments = Q. To what extent do you agree that Integrated Reporting promotes a more connected and efficient approach and could replace other forms of reporting approach in India?

1.2.5 Limitations

The study is based on secondary data, available literature and stakeholders' perceptions. As corporate reporting for ESG disclosure is continuously evolving, the period of survey conducted to gather perception on the <IR> framework has become one of the limitations of this study. Secondly, the perception is limited to Indian stakeholders.

1.3 Result Analysis

1.3.1 RQ 1: What Are the Limitations of today's Corporate Reporting System in India?

Before investigating the efficiency of the <IR> framework, it's important to study the limitations of today's corporate reporting system. Hence, in this question, twelve limitations have been provided to the participants with multi-check options.

The results (Fig. 1.2) reveal that 57% of responders agreed that there are '***too many frameworks***' in India for corporate reporting that lead to '***too many initiatives***' (54%) with '***insufficient focus on medium and long-term***' disclosures (50%).

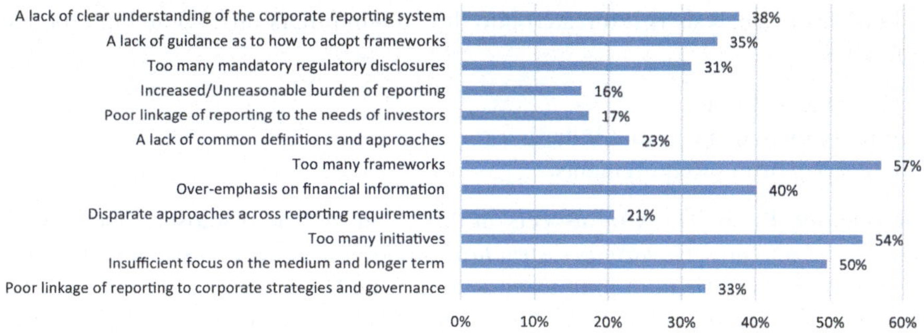

Fig. 1.2 Which of the following concerns do you think apply to today's Corporate reporting system? *Source* Primary Data

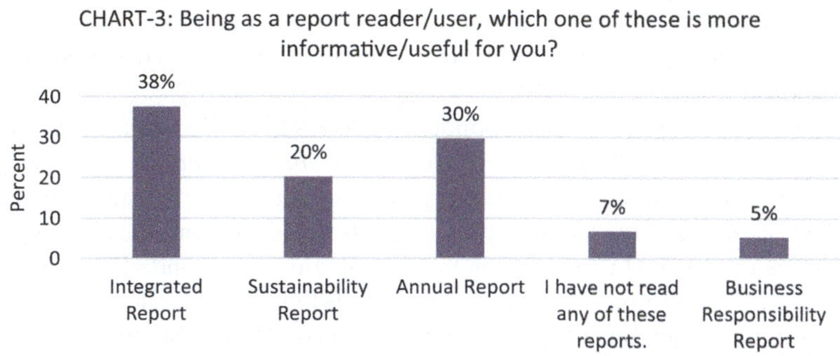

Fig. 1.3 Being a report reader/user, which one of these is more informative/useful for you? *Source* Primary data

1.3.2 RQ 2: How Effective is the <IR> Framework in ESG Disclosure?

The efficiency of any reporting framework depends on how it is effective in communicating the value creation process, strategy and holistic materiality elements of an organisation as per the stakeholders' needs. Therefore, the perception of stakeholders has been gathered and analysed to understand the effectiveness of <IR> Framework in meeting those needs.

As there are multiple reporting frameworks prevailing in India, firstly, it is important to understand which one of the reports is more effective and useful. Thus, the findings (Fig. 1.3) reveal that 38% of respondents believed that the integrated report is more informative and useful to them as compared to the sustainability report (which is generally based on GRI); regular Annual Report and BRR.[1]

[1] BRR now known as BRSR (Business Responsibility and Sustainability Report).

Further our analysis is bifurcated into three main pillars of ESG and interpreted as follows:

- Governance Pillar
- Environmental and Social Pillar
- Other Reporting Requirements

Governance Pillar The vital aspects in this pillar are the disclosure about *Business Model* and how *Governance structure* supports its ability to sustainably value creation over short-medium-long-term; secondly the *Materiality Matters* and third is organisations' *Strategic* disclosures that would meet the expectation of *Stakeholders and its Engagement.*

The study reveals (Fig. 1.4) that 56% of respondents believe that <IR> is '*Extremely Effective*' in disclosing organisations' *business model as well as its value creation process*, whereas 34% of respondents believe that it is just effective. However, only 47% of respondents agreed that <IR> is '*Extremely Effective*' in disclosing governance structure ability and the materiality matters to the organisation (46% response).

It is also agreed by 52% of respondents that <IR> is '*Extremely Effective*' in disclosing the *Risks and Opportunities and how the governance is dealing with them.* The efficiency of <IR> in providing insight about the *organisation's strategy* was 46% '*Extremely Effective*' and providing information about the challenges and uncertainties incurred while executing those strategies was found 50% '*Effective*'.

Thus, it can be determined from the above results that <IR> is efficient in providing information that would meet the requirements of G-Pillar of ESG to some extent.

Environmental and Social Pillar. The disclosure in the environmental and social pillar is the outcome of *Materiality Matters* to the organisation and varies sector by sector. Hence, the level of questions has been kept broad and mainly categorised into five questions as shown in chart-4 below:

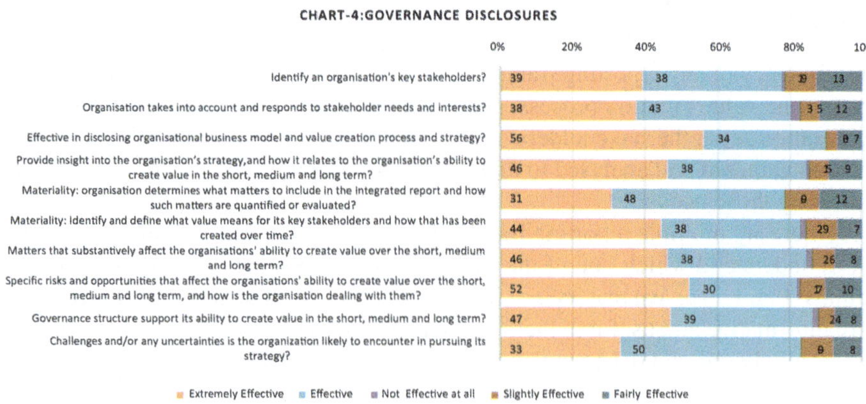

Fig. 1.4 Governance disclosure. *Source* Primary data

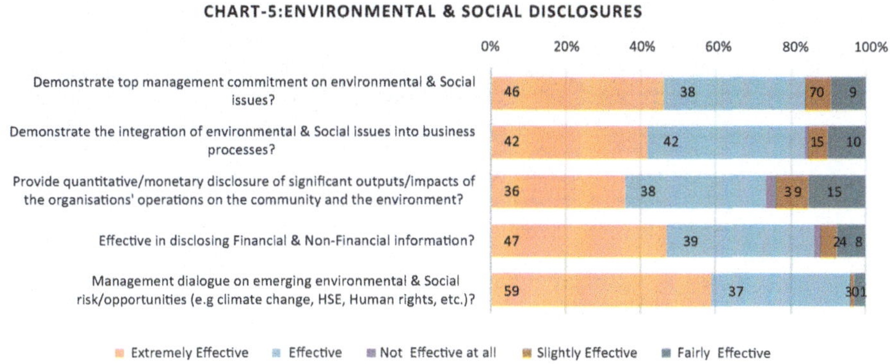

Fig. 1.5 Environmental and Social disclosures. *Source* Primary data

It can be observed from Fig. 1.5 that the response rate for *'Extremely Effective' and 'Effective' is quite similar.* The study reveals that for demonstrating the top management commitments on environmental and social issues and disclosing the non-financial and financial information, the response rate is quite similar, i.e. 46–47% agreed that <IR> is '*Extremely Effective*'. Whereas 59% of respondents find <IR> as '*Extremely Effective*' in disclosing management dialogue on emerging environmental and social risks/opportunities. However, only 36% believe that <IR> is '*Extremely Effective*' in providing quantitative/monetary disclosure of significant outputs/impacts of the organisations' operations on the community and the environment.

Thus, it can be determined from the above results that as compared to the G pillar, <IR> is more efficient in providing information that would meet the requirements of the E-S Pillar of ESG.

Other Reporting Requirements. Besides the material topics in all the three pillars of ESG, there are other reporting requirements that make the ESG reporting more insightful, understandable and holistic in nature such as providing future-oriented information either quantitative/qualitative to the stakeholders with short-medium-long-term targets, showcasing trends, including benchmarking context, etc. and maintaining the decent level of transparency by disclosing both positive/negative aspects in overall value creation process, nonetheless should be accurate, reliable and presented on regular basis.

Hence, the above mentioned requirements have been administered to the participants and the responses are summarised in Fig. 1.6.

The analysis reveals that 62% agreed that the reporting should be done annually, 50% and 51% agreed that <IR> is '*Extremely Effective*' in disclosing short-medium-long-term targets and showcasing trends, respectively. While 48% agreed that <IR> is '*Extremely Effective*' in showing a holistic picture of the combination, interrelatedness and dependencies between the factors that affect the organisations' ability to

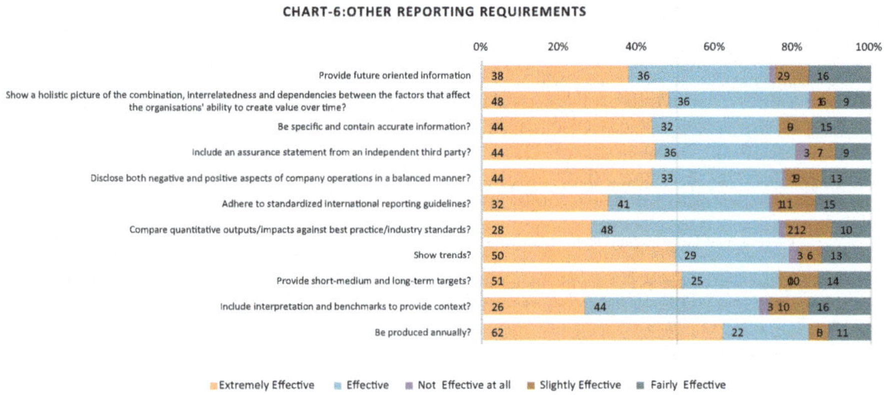

Fig. 1.6 Other reporting requirements *Source* Primary data

create value over time. Despite that fact, 44% found <IR> to be '*Extremely Effective*' in maintaining transparency by disclosing both positive/negative aspects, and providing third-party independent assurance statements to uphold the reliability and accuracy of the information.

Thus, it can be determined from the above results that <IR> is effective in meeting the other requirements of ESG disclosures.

1.3.3 *Is <IR> framework Efficient to Replace the Sustainability Report, Annual Report and BRR?*

We collected participants' sentiments to study the capability of <IR> framework in replacing other reporting approaches. The results revealed that 64% of respondents (the companies who have adopted <IR> , N = 76) '*Agreed*' and 18% '*Strongly Agreed*' that <IR> framework is efficient in replacing other reporting approaches. To test the hypothesis, a contingency table was prepared with two variables, viz. IR_Replace_Sentiments and IR_Umbrella_Framework_Sentiments in SPSS with Pearson Chi-Square ($\chi^{2)}$ and Cramer's V ($\varphi_{c)}$. The results (w.r.t. Table 1.2) reveal that there is association with these two variables significantly at χ^2 (DF = 4) is 64.777 within p-value ($<0.000 < \alpha = 0.05$) and the value of $\varphi_c = 0.653$ indicates strong association (w.r.t. Table 1.1). Thus, we accepted H_1. However, there is a warning of expected cell count as given by Cochran (1954) which seems to be despairing to Agresti (1990) to cover all cases within one rule, and therefore, we extended our analysis with Monte Carlo p-values (Mehta and Patel 2010) though both the methods yielded same p-values. Thus, it is concluded that <IR> is an 'umbrella' for corporate reporting and could replace other forms of reporting approach in India (Table 1.2).

Table 1.1 Symmetric measures

		Value	Approximate Significance	Monte carlo significance		
				Significance	99% Confidence Interval	
					Lower bound	Upper bound
Nominal by nominal	Phi	0.923	0.000	0.000[c]	0.000	0.000
	Cramer's V	0.653	0.000	0.000[c]	0.000	0.000
N of valid cases		76				

c. Based on 10,000 sampled tables with starting seed 2,000,000

Source SPSS output (Researche"s Compilation)

1.4 Summary and Conclusion

In an endeavour to discover *One Planet One compliance?* It was discovered that there is a diverse reporting framework for one ESG and each has its own efficiency like TCFD and CDSB possess its own uniqueness and is highly efficient for disclosing climate-related issues. While, GRI and SASB provides standard & metrics across all the three pillars of ESG. Besides, <IR> is efficient in providing guidance and principles on 'what to be reported and to what extent'. Additionally, from the stakeholder perception analysis it was revealed that there is scientific evidence [χ^2 (df = 4) is 64.777 with p-value ($< 0.000 < \alpha = 0.05$)] and the value of $\varphi_c = 0.653$) that <IR> could replace other reporting approaches such as Sustainability Report, Annual Report and BRR as it is '*Extremely Effective*' in disclosing Business model, its value creation process, Risk and Opportunities, Management dialogue on Management dialogue on emerging environmental & Social risk/opportunities (e.g. climate change, HSE, Human rights, etc.). Thus, a collaboration of these organisations under one umbrella of Value Reporting Foundation and subsequently merging with IFRS Foundation is a pathway towards *One Planet One compliance.*

1.5 Recommendation and Direction for Future Research

The stock exchange board should think about mandating the <IR> framework for the sectors that are highly environmentally sensitive, however, adopting this framework voluntarily by companies will benefit them in demonstrating transparency and accountability towards a sustainable economy.

An in-depth study on the efficiency of the <IR> framework could be conducted by exploring the material topics of ESG, e.g. how efficient is this framework in disclosing Human Rights issues, Occupational Health and Safety, Environmental metrics, Corporate Governance practices required for social benefits and environmental protection. By doing so, it would be supportive of extending our proposed

Table 1.2 Chi-square tests

	Value	df	Asymptotic significance (2-sided)	Monte Carlo Sig. (2-sided)			Monte Carlo Sig. (1-sided)		
				Significance	99% Confidence interval		Significance	99% Confidence interval	
					Lower bound	Upper bound		Lower bound	Upper bound
Pearson Chi-Square	64.777[a]	4	0.000	0.000[b]	0.000	0.000			
Likelihood ratio	49.815	4	0.000	0.000[b]	0.000	0.000			
Fisher-Freeman-Halton exact test	43.781			0.000[b]	0.000	0.000			
Linear-by-linear association	35.020c	1	0.000	0.000[b]	0.000	0.000	0.000[b]	0.000	0.000
N of valid cases	76								

a. 5 cells (55.6%) have an expected count of less than 5. The minimum expected count is 0.66

b. Based on 10,000 sampled tables with starting seed 2,000,000

c. The standardised statistic is 5.918

Source SPSS output (Researcher's Compilation)

model. Another avenue could be exploring the interrelatedness of GRI, SASB, IR, TCFD and CDP with empirical analysis.

References

Agresti A (1990) Categorical data analysis. Wiley, New York

Ahmed Haji A, Anifowose M (2017) Initial trends in corporate disclosures following the introduction of integrated reporting practice in South Africa. J Intellect Cap 373–399.https://doi.org/10.1108/JIC-01-2016-0020

Ara M, Harani B, D (2020a) Integrated reporting insight: why organisation voluntarily reports? Int J Sci Technol Res 9(1):3055–3069

Baron R (2014) The evolution of corporate reporting for integrated performance. Background paper for the 30th round table on sustainable development (pp 1–36). OECD, Paris. https://www.researchgate.net/profile/Richard-Baron-2/publication/275038287_The_Evolution_of_Corporate_Reporting_for_an_Integrated_Performance/links/56714ef308aececfd5551f94/The-Evolution-of-Corporate-Reporting-for-an-Integrated-Performance.pdf?origin=publ

Bartha M, Cahan S, Chen L, Venter E (2017) The economic consequences associated with integrated report quality: early evidence from a mandatory setting. SSRN, 1–47. https://www.rwp.bwl.uni-muenchen.de/forschung/wiss_veranstaltungen/arw---neu/paper_arw-marybarth_09_05_16.pdf

Bernardi C, Stark A (2016) Environmental, social and governance disclosure, integrated reporting, and the accuracy of analyst forecasts. Br Account Rev 1–52.https://doi.org/10.1016/j.bar.2016.10.001

CDP (2023a) What we do. CDP. https://www.cdp.net/en/info/about-us/what-we-do

CDP (2023b) Why disclose as a company. CDP. https://www.cdp.net/en/companies-discloser

CDP (2023c) States and regions. CDP. https://www.cdp.net/en/states-and-regions

CDSB (2022) CDSB framework for reporting environmental & social information. CDSB, United Kingdom. https://www.cdsb.net/sites/default/files/cdsb_framework_2022.pdf

CDSB (2023) Our story. CDSB. https://www.cdsb.net/our-story

Cochran WG (1954) Some methods of strengthening the common chi-square tests. Biometrics 417–451. https://doi.org/10.2307/3001616

Corvino A, Doni F, Martini S (2020) Corporate governance, integrated reporting and environmental disclosure: evidence from the South African Context. Sustainability 1–19.https://doi.org/10.3390/su12124820

Cosmulese CG, Socoliuc M, Ciubotariu MS, Mihaila S, Grosu V (2019) An empirical analysis of stakeholders' expectations and integrated reporting quality. Economic Research-Ekonomska Istraživanja, 3963–3986. https://doi.org/10.1080/1331677X.2019.1680303

Gelmini L, Vola P (2021) Integrated reporting and environmental disclosure: is natural capital neglected? Corp Ownersh Control 18(2):131–139. https://virtusinterpress.org/IMG/pdf/cocv18i2art10.pdf

GRI (2023) Our mission and history. GRI. https://www.globalreporting.org/about-gri/mission-history/

Haruna M, Mkhize M (2020) Integrated reporting practice in a developing country—Ghana: legitimacy or stakeholder oriented? Int J Discl Gov 1–15.https://doi.org/10.1057/s41310-020-00092-z

Hoang T (2018) The role of the integrated reporting in raising awareness of environmental, social and corporate governance (ESG) performance. Stakehold Gov Responsib 14:47–69. https://doi.org/10.1108/S2043-052320180000014003

Hoque ME (2017) Why company should adopt integrated reporting? Int J Econ Financ Issues 7(1):241–248. https://dergipark.org.tr/download/article-file/364169

IFRS (2021) IFRS foundation announces international sustainability standards board, consolidation with CDSB and VRF, and publication of prototype disclosure requirements. IFRS. https://www.ifrs.org/news-and-events/news/2021/11/ifrs-foundation-announces-issb-consolidation-with-cdsb-vrf-publication-of-prototypes/

IIRC (2013) The international <IR> framework. IIRC, London. https://integratedreporting.org/integratedreport2017/pdf/IIRC_IR_Framework.pdf

IIRC (2019a) Evidence that the 'momentum phase' is taking off. IIRC. https://integratedreporting.org/news/evidence-that-the-momentum-phase-is-taking-off/

IIRC (2020) 10 years of inspiring integrated thinking and reporting. IIRC. https://integratedreporting.org/10-years/

IIRC (2021) International <IR> framework. IIRC, London. https://integratedreporting.org/wp-content/uploads/2021/01/InternationalIntegratedReportingFramework.pdf

IIRC (2022) The growing momentum for integrated reporting: part 1. IIRC. https://www.integratedreporting.org/news/the-growing-momentum-for-integrated-reporting-part-1/

KPMG (2022) Big shifts, small steps: survey of sustainability reporting. KPMG International. https://assets.kpmg.com/content/dam/kpmg/xx/pdf/2022/10/ssr-small-steps-big-shifts.pdf

Lee K, Yeo G-H (2015) The association between integrated reporting and firm. Rev Quant Finance Account 1221–1250.https://doi.org/10.1007/s11156-015-

Mähönen J (2020) Integrated reporting and sustainable corporate governance from european perspective. Accounting, economics, and law: a convivium, pp 1–40. https://doi.org/10.1515/ael-2018-0048

Martínez-Ferrero J, García-Sánche I, Ruiz-Cano D (2016) The casual link between sustainable disclosure and information asymmetry: the moderating role of the stakeholder protection context. Corp Soc Responsib Environ Manag 319–332

Mehta C, Patel N (2010) IBM SPSS exact tests. IBM SPSS, Cambridge, Massachusetts. https://www.researchgate.net/profile/Cyrus-Mehta/publication/265357333_SPSS_exact_tests/links/54aaac490cf25c4c472f45e7/SPSS-exact-tests.pdf?origin=publicationDetail&_sg%5B0%5D=Skbc_CnxGc8xsSHOURD3l6j1K3vNj9AR2F-NT7sq2S2s6XVi_HTzH2-eU6N5yAHftoH_g_hgAFOx9Mc

Ofoegbu GN, Odoemelam N, Okafo R (2018) Corporate board characteristics and environmentaldisclosure quantity: evidence from South Africa(integrated reporting) and Nigeria (traditionalreporting). Cogent Bus Manag 1–27. https://www.tandfonline.com/doi/epdf/10.1080/23311975.2018.1551510?needAccess=true&role=button

Park SR, Oh K-S (2022) Integration of ESG information into individual investors' corporate investment decisions: utilizing the UTAUT framework. Front Psychol. https://doi.org/10.3389/fpsyg.2022.899480

Raimo N, Nuccio E, Vitolla F (2021) Corporate governance and environmental disclosure through integrated reporting. Meas Bus Excell 26(4):451–470. https://doi.org/10.1108/MBE-05-2021-0066

SASB (2023) SASB Standards connect business and investors on the financial impacts of sustainability. SASB. https://www.sasb.org/about/

SASB (2023b) About us. SASB. https://www.sasb.org/about/

Setyawan H, Kamilla P (2015) Impact of corporate governance on corporate environmental disclosure: Indonesian evidence. In: International conference on trends in economics, humanities and management (ICTEHM'15) August 12–13, Pattaya (Thailand), (pp 1–6). http://icehm.org/upload/2392ED815026.pdf

Singaraju S, Niininen O, Sullivan-Mort G, Nguyen Q (2016) Social media and value co-creation in multi-stakeholder systems: a resource integration approach. Ind Mark Manage 54:44–55

Sridharan V (2018) Bridging the disclosure gap: investor perspectives on environmental, social & governance (ESG) disclosures. University of Pennsylvania, Department of Earth and Environmental Science. University of Pennsylvania ScholarlyCommons. https://repository.upenn.edu/mes_capstones/72

TCFD (2022) TCFD status report. TCFD, Switzerland. https://assets.bbhub.io/company/sites/60/2022/10/2022-TCFD-Status-Report.pdf

Umoren AO, Udo E, George B (2015) Environmental, social and governance disclosures: a call. J Finance Account 227–233.https://doi.org/10.11648/j.jfa.20150306.19

UNCTAD (2012) Report of the intergovernmental working group of experts on international standards of accounting and reporting on its twenty-ninth session. UNCTAD, Geneva

UNCTAD (2014) Best practice guidance for policymakers and stock exchanges on sustainability reporting initiatives. New York and Geneva: sustainability reporting initiatives. https://unctad.org/system/files/official-document/diaeed2013d6_en.pdf

UNFCCC (2015) The Paris agreement. UNFCCC. http://unfccc.int/files/essential_background/convention/application/pdf/english_paris_agreement.pdf

UNSSE (2015) Model guidance on reporting ESG information to investors: a voluntary tool for stock exchanges to guide issuers. United Nations Sustainable Stock Exchanges (SSE) initiative. https://d306pr3pise04h.cloudfront.net/docs/issues_doc%2FFinancial_markets%2FSSE-model-guidance-on-reporting-ESG.pdf

UNSSE (2023) ESG disclosure guidance database. Retrieved 2023, from Unite Nation sustainability stock exchange initiative. https://sseinitiative.org/esg-guidance-database/

Zhou S (2017) Does integrated reporting matter to the capital market? Abacus 53(1):94–132. https://onlinelibrary.wiley.com/doi/epdf/10.1111/abac.12104

Chapter 2
Sustainable Influence of Instagram and Its Linkages on Gen Z's Purchase Intention

Anitha Nallasivam and S. Mahalakshmi

Abstract Covid-19 brought in a lot of new technological developments when it plagued the world. Countries across stayed connected via social media. Applications such as Instagram, TikTok, and YouTube were ways to stay connected. Social media started off as a mere platform to connect but over the years it has transformed into so much more. Instagram has grown to be a marketplace. Businesses now have increased visibility and a better chance to reach their target audience. Gen Z particularly used such platforms for staying connected, for education, and even for entertainment. Instagram is the third most used application in the world and it has opened an array of opportunities both to the seller and to the buyer. A decade ago, no one would have thought that Instagram would end up being a marketplace too. India has one of the biggest Gen Z populations. Which is approx. 250 million people. Giving marketers a very new target audience generation, one that is thoughtful about a lot of agendas. Such as sustainability and ethics. Today's customer has become demanding and it's more difficult to convert them. But digital marketing makes it easier to make the right buyer and seller meet. In this research, we discover how Gen Z's purchase behavior is on Instagram. With that intention, the study topic was chosen for research as it is relevant to today's time and marketing practices. the study is descriptive in nature. A sample size of 86 was chosen randomly from the Gen Z population and data was collected using the survey method called structured questionnaire tool. Collected data was analyzed, interpreted, and presented in the report for further extension of research.

Keywords Generation Z · Target audience · Digital marketing sustainability · Purchase behavior · Instagram

---

A. Nallasivam · S. Mahalakshmi (✉)
Jain Deemed-to-be University, Bangalore, India
e-mail: dr.mahalakshmibharath@gmail.com

A. Nallasivam
e-mail: dr.anithanallasivam@cms.ac.in

A. Edward J. et al. (eds.), *Digital Transformation for Business Sustainability*, Contributions to Environmental Sciences & Innovative Business Technology,
https://doi.org/10.1007/978-981-99-7058-2_2

2.1 Introduction

Instagram keeps the consumer informed about trending products. It's a perfect tool for competitor analysis and gives customers a variety to choose from. With everything just being a click away, it's easier to keep tabs on the competition. For the companies, one must note that by using tools such as social blade, one can assess a lot through just an Instagram handle. All statistics of engagement, money earned, new followers, average comment rate, etc., can be determined (Agrawal 2022).

In terms of brand visibility, Instagram marketing is fantastic. According to survey participants, Instagram's posts and reels make it easier for them to find new goods or services (Sun et al. 2022). Due to the extension of latitudinal interaction with a business on Instagram, people have greater faith in the brands that are promoted there. In contrast, multiple studies also discovered that firms without an Instagram account are more likely to be ignored in the present and the future by potential customers. A brand can benefit much from enhancing its standing and clearing the air. Fashion Nova, Lush, Lea Clothes, and other companies have made significant use of social networking. Some companies have successfully displayed their product categories and gained customers.

One of the things to notice is that the above-mentioned brands are not just merely advertising but creating a brand community. These communities have become stronger after the Covid-19 era as many users followed their favorite brands closely for updates and other relevant information. It's not just enough to be present on Instagram. But there is an algorithm to understand (Bellaaj 2021). The base of it includes engaging with your followers, posting quality content that is on par with the trends on the internet, and maintaining a balance using tools such as Instagram stories, posts, reels, guides, and shops. Previously, content creators could put up longer videos on IGTV, but now the IGTV format has been discontinued.

Instagram over the years has a trial period for each of the functions that it has to offer. To encourage brands and influencers to continuously upload quality content, Instagram now offers monetary benefits. The benefits are achieved when there is a continuous effort put into delivering quality content. A customer experience is extremely hyped up once they are noticed, Gen Z are often seen engaging with their favorite brands and influencers.

According to HubSpot (Instagram engagement report, 2020) where 81,258,182 In order to comprehend a number of criteria, Instagram posts and 1,676,748 of the top users were being monitored. Engagement, influencers, and caption length are a few important findings that have considerable significance for any company looking to sell itself successfully. The following are some of the study's major highlights: More people liked reels and photo posts than permanent video posts on a page, 27.55% more. On Instagram, a post includes 10.7 hashtags on average, although 50% of posts have fewer than 6 hashtags. The pages devoted to beauty and cosmetics receive the highest engagement. 52.3% of Instagram users follow fewer than 1000 additional users. One of Instagram's most fascinating features is the geography tag, which lets your followers know where you are (Król and Zdonek 2021).

Below is a list of compelling statistics by HubSpot that makes Instagram marketing and purchase decisions of Gen Z, a brilliant topic of study for the current market environment, especially post covid-19. Screen time has increased and so have consumption patterns. These are some statistics to keep in mind:

- Instagram has over one billion active users.
- 64% of the users are between the ages of 16–34.
- 52% of Instagram users are female and 48% male.
- Influencer marketing grew by 48% in FY 2020–2021.
- There are 8 million plus business accounts.
- 84% of the Instagram influencers are female.
- 7 out of 10 hashtags are branded keywords.
- 500 million users use Instagram stories every day.

After going through the facts, one thinks that only a beauty or a lifestyle brand has a chance for recognition. This is highly untrue; there are new trends emerging for every kind of customer out there. Knowing your customers is now simpler than ever, and data collecting has been quicker. Businesses may use Instagram to connect with the right customers and experience exponential brand growth. Whether it's a B2B corporation or a charity (Baykal 2020). A brand must, however, be aware of the shifting trends and how they may affect its product marketing plan.

Many businesses find it suitable to promote on Instagram. The estimated number of advertisers per month is around 2 million advertisers on Instagram. the approximate audience that is present is about 1.32 billion users. This means that it roughly reaches about 23% of the global population aged 13 and older. 48% of Instagram users are observed to be men and 52% are women. Instagram is the 7th most visited website and the 9th most googled term. It is also the fourth most used social media followed by WhatsApp, YouTube, and Facebook. It has become the second most downloaded app in the world and there are many reasons that add up to making it such a desirable platform to sell and advertise (Dorie and Loranger 2020). Therefore, it becomes very important to discover how Gen Z's purchase behavior is on Instagram relevant to today's time and marketing practices, Gen Z was the population taken for the study, Primary data was collected from the respondents using the structured questionnaire using Likert scale and secondary data was collected from various publications web pages, books, and newspapers. The hypothesis was formulated to address the research questions and specific statistical tools like regression and ANOVA tools were used to analyze the data.

The Present study is structured as follows.

2.2 Statement of the Research Problem

The problem forms a triggering force behind any research work. In more recent times, there has been an increase in the way Gen Z consumes technology. The pandemic (Covid-19) pushed innovation in terms of technology and product development.

Many believe that this period has made people more tech-savvy but has introduced a problem of overbrowsing, but it is a great opportunity for marketers to get noticed. There is a huge shift from offline to online markets that needs to be studied. There are multiple avenues available to today's customers. It's not like the earlier times when it would be impossible to address grievances, shop from non-traditional platforms, and discover unique forms of promotion or activities. In today's world, it takes a lot to convert a customer. And Generation Z is not easy to please, it takes more than just normal advertisements to convert them into potential customers. The way we interact on Instagram has completely changed. The worth of a product is challenged and creative ways are used by businesses and Instagram influencers to understand their audience. Traditional marketers have now realized that one channel is not enough to achieve success (Hameed and Mathur 2020). There are many aspects to take care of on Instagram in order to be a visible brand for, e.g., trending hashtags, influencer genre, and high engagement rates are some things that brands extensively research. This research aims to understand Instagram influencers and how they affect the purchase behavior of Gen Z. It is observed that there are various marketing strategies for various groups. In the past, there haven't been many studies that have been conducted on GenZ purchase behavior when exposed to social media. It is a field that can be discovered further and make a clear decision for marketing managers whether or not to use Instagram marketing as a tool to promote their business. There is a gap in understanding the value of brand engagement, the use of micro-celebrities, and influencer marketing which has to be understood before coming up with an annual marketing budget (Herrando et al. 2019).

2.3 Review of Literature

Agarwal (2022) summarized that at the end of the twentieth century, many things altered as a result of the advent of digital technology, including people's psyche. The buying and shopping interests of consumers have changed over time. The Internet, smartphone-led digital technology, and mobile applications have caused a sizable portion of the population to switch from traditional markets to online channels. More information is given to the buyer in this way than at a physical store. The newest and hottest trends have an impact on people. Shoppers who are tech-savvy are members of the Y and Z generations, who were born between 1981 and 1885. An examination of generational cohort-based buying provides valuable insight into consumer motivations and online purchase involvement that were influenced by shared values and beliefs.

Sun et al. (2022) studied that China is a sizable market that accounts for one-third of the worldwide market for luxury goods. By 2025, Chinese consumers are expected to contribute RMB 1 trillion to global luxury sales according to further research into the buying patterns of Gen Z consumers. Additionally, since the 1990s, digital marketing and internet sales have surpassed traditional ones due to technical innovation. Social media has a significant impact on retail trends and consumer behavior. By 2025, internet contact will have an impact on the consumption of luxury goods. Luxury companies and stores are motivated by this to comprehend the demand created by Instagram alone. Brands now place a high value on a Gen Z user's digital presence. Instagram is assisting on a global scale.

Bellaj (2021) studied that e-commerce presence is increasing a lot in today's environment. A digital channel is known as a digital shopping format that is provided to consumers by retailers to give them an online shopping experience. Digital Channels have transformed how marketing activities take place and how to manage customer relationships, especially for online businesses. Entrepreneurs now apply digital technologies to increase aspects of their business activities. As reported recently, more than half of the world's population now uses social media and there are more than 150 million businesses that use Instagram, WhatsApp, and Facebook. More than 100 million businesses interact with their customers via Instagram. Digital channels help businesses collect and manage data. It's easier to connect information about the customer now. Social media is now used for creating, strengthening, and enlarging business networks. Entrepreneurs use digital networks to enhance their brand visibility, improve relationships with their existing and future customers, and expand their market share. Digital channels help a business generate value and improve business performance. Reports suggest that digital channels act as a backbone for businesses in emerging economies.

Krol and Zdonek (2021) studied social media use and its impact on intrinsic motivation in Generation Z. Gen Z is a very different type of target audience when compared to other audiences. The same kind of advertising efforts that were done before are not very applicable to today's audiences. Mobile phones have become the number one place to receive content. ICT information and communication innovations have increased the experience of online organizations. It is a time where every organization should optimize their resources using modern solutions which brought a revolution to business market communication. Social media is a contributing factor to this process. Information is ready and accessible in today's time. Gen Z interacts I'm a certain way, they stay connected via the internet, and social media for communication purposes. This consumer group is active in nature, who with the help of online platforms share information, look for entertainment, etc. Therefore, in this study, it was important to understand the frequency with which social media platforms are used and how they are used as a tool for communication in the years to come.

Dorie and Loranger (2020) studied the multi-generation differences in channel activity. The retail market has become increasingly competitive, retailers now realize that there is an urgent need to connect with the customers through a wide range of channels that act as touchpoints between the retailer and the customer. Many retailers want to research and understand how shopping patterns vary. It's not just about Brick-and-Mortar stores, the younger generations are riding a wave of online shopping. Many customers use mixed channels for their shopping interactions but there is a significant growth in online channels. When a customer gets the freedom to interact via social media, phones, and tablets they get a better insight on their choice of brand. Pew Research (2018) has found that Generation Z, millennials, and baby boomers owned smartphones at rates 92,85 and 67%. Instagram has a total of one billion active users making it an engaging platform for both businesses and users. There has been a rise in the number of channels available to a customer, and hence it has become increasingly important for a marketer to understand the multichannel behavior.

According to Baykal (2020), these days, there is a massive change in which the marketplace functions. Customers look for improved experiences where it's fully focused on them. The marketplace has become more responsive to customer needs. The omnichannel concept is a way of blurring lines between the online and physical stores. Customers look for a mix of both omnichannel environments, they have a choice to move freely among online and physical stores and expect a level of professional seamlessness. After looking at various generations, it was found that GenZ deserves a separate marketing campaign. These include the rise of omnichannel experiences. Technology has advanced how the retail sector functions. Consumers are not just limited to the product, but also look at information from different sources and often use a pattern of multiple channels for shopping, stores, and browsing the internet.

Hameed and Mathur (2020) researched that India has the second-highest population on the entire planet. The greatest Gen Z population in the world is estimated to be 472 million people, according to Hameed and Mathur (Hameed and Mathur 2021). The distribution of Gen Z's demographics has changed. They are more politically and socially conscious, have new job objectives and prospects, and have distinct workplace preferences. Also, as a result of these features, consumption is altering. Gen Z has distinct thoughts and ideas about how they might contribute to the economy of India, according to a study that used focus groups to analyze how consumption patterns differ.

Herrando et al. (2019) studied that social media commerce has gained a lot of attention lately. Social commerce websites have a unique feature where there can be an exchange of information among various customers. This kind of system is a mix of social media, e-commerce, and social networks. They receive information from trustworthy sources. Websites generate information for a company and help them to influence other users' purchase decisions. Consumers these days depend on online information for better-informed decisions. Product reviews have become extremely important to assess the quality beforehand. 78% of online users are influenced by social media and many were engaged with feedback and reviews. The kind of purchase that takes place in social commerce is different from regular e-commerce. This reason for differentiation is because of the participation aspect since users can generate content and share their feedback. The purchase intention of a Gen Z consumer is based on hedonic, utilitarian, and social values. Social media marketing has flourished because of the expansion of social media. (Instagram, Facebook, Twitter making customers feel heard. Brands now have a greater chance to reach their target customers. Brands have a multichannel way of promoting special offers. They advertise with their own media pages but also hire and collaborate with influencers to promote products and special discounts. Social media has become popular after some creators found a substantial number of followers by producing content on a daily basis (lifestyle, technology, beauty, and fashion) which makes their fans relate to them better.

2.4 Research Gaps

Our potential research gap is the studies done on Instagram's Influence and purchase decision of Gen Z. There are no long-term studies that have been done in order to understand the two. Many existing studies have focused on business aspects, luxury marketing, and tools used by Instagram. But there is a rise in the need to understand why there is a need to study a specific generation, their buying habits, and brand exposure on Instagram. This could help businesses on Instagram understand the buying behavior of their target audience.

Another gap would be in the relationship of the location of the customer and their buying behavior. It is important to understand how urban and rural area customers are different from one another when it comes to purchasing habits. There are very few studies done on GenZ that stay in Rural areas or Tier 4 cities of India. Many businesses and influencers struggle to have a metric or a way to measure the success of a social media campaign. A lot of concepts such as bounce rate, significant amount of time spent browsing, and leads can't be accounted for.

2.5 Research Objectives

To study the demographic profile of the respondents and study similarities in the way they interact with Instagram.

To understand the factors that influence the purchasing behavior of Generation Z (Age 8–23).

To know about how the brands are leveraging social media to increase brand awareness and increase sales conversion.

To understand the increased consumption and interaction on Instagram post Covid-19 and dependency on technology.

2.6 Methodology

The research is descriptive in nature. Gen Z was the population taken for the study and Eighty-Six Gen Z samples were randomly chosen and reprinting the entire population. Primary as well as secondary data sources were used to collect the data. Primary data was collected from the respondents by the structured questionnaire using the Likert scale, and secondary data was collected from various publications' web pages, books, and newspapers. The hypothesis was formulated to address the research questions and specific statistical tools like regression and ANOVA tools were used to analyze the collected data, analyzed, interpreted, and presented in the report as research findings.

Hypothesis 1

H0: There is no significant association between the type of product and the channel used.

H1: There is a significant association between the type of product and the channel used.

Hypothesis 2

H0: There is no significant relationship between time spent on Instagram based on gender, age, and location.

H1: There is a significant relationship between time spent on Instagram based on gender, age, and location.

CHI-SQUARE TEST: The chi-square test, a statistical method, is used to compare actual results to forecasts. This test seeks to establish if a difference between actual and projected data is the result of chance or a relationship between the variables under investigation. As a result, the chi-square test is a fantastic tool for assisting us in understanding and assessing the relationship between our two category variables.

A chi-square test or similar nonparametric test must be used to examine the distribution of a categorical variable. Categories like nations or animals can be indicated by nominal or ordinal categorical variables. Since they can only have a few particular values, they cannot have a normal distribution. For example, a marketing firm in India wants to find the relationship between the link the product type purchased and the Instagram channel used to gain awareness. It is mainly used to calculate two categorical variables, which are:

- As a result of chance
- Because they have a relationship between them

The formula for Chi-Square $X2 = \sum (O - E)2 / E$.

where C = Degree of freedom, O = Observed value, and E = Expected value. The number of variables that can change in a calculation is represented by the degrees of freedom in a statistical calculation. It is possible to determine the degrees of freedom to make sure chi-square tests are statistically reliable. These tests are typically employed to contrast the observed data with the data that would be anticipated if a specific hypothesis were to be correct. The values you compile on your own make up the observed values. Based on the null hypothesis, the predicted values are the anticipated frequencies.

A statistical test called chi-square examines how categorical variables from a random sample differ from one another to determine whether the results are consistent with what was anticipated. The following circumstances call for the application of the Chi-Squared test:

The Chi-squared test can be used to check whether your data resembles a well-known theoretical probability distribution, such as the Normal or Poisson distribution.

The Chi-squared test can be used to assess the goodness of fit tested regression model on the training, validation, and test data sets. The chi-squared test is the procedure used the most frequently to evaluate hypotheses. A claim that any given scenario might be true can be tested later and is known as a hypothesis. The Chi-Square test determines how much of a difference there is between the expected and actual results when the sample size and the number of variables in the relationship are given.

These tests use degrees of freedom to determine if a certain null hypothesis can be rejected based on the total number of observations made throughout the experiment. The larger the sample size, the more dependable the result.

To understand the dependency of place and selection of brands, the Observed Values and expected values are shown in Tables 2.1 and 2.2.

X 2 = 6.4178.

Number of Rows = 5.

Number of Columns = 5.

Degree of Freedom = 16.

The calculated value is more than 0.5 so the null hypothesis will be rejected. So there is an association between the type pf product and the channel used.

Hypothesis 2 H0: There are no significant effects on time spent on Instagram based on gender, age, and location.

Table 2.1 Observed values

	Brand ads	Reels	E commerce links	Creator post	Trending terms	Total
Clothes	6	2	3	5	2	18
Footwear	4	2	5	5	3	19
Electronics	4	1	6	3	4	18
Travel packages	5	2	3	4	2	16
Food	3	3	2	4	3	15
Total	23	10	19	21	14	86

Source Primary Data

Table 2.2 Expected values

	Brand ads	Reels	E commerce links	Creator post	Trending terms	Total
Clothes	4.6	2.09	3.97	4.39	2.93	18
Footwear	4.86	2.2	4.19	4.63	3.09	19
Electronics	4.6	2.09	3.97	4.39	2.93	18
Travel packages	4.09	1.86	3.53	3.9	2.6	16
Food	3.83	1.74	3.31	3.66	2.44	15
Total	22	10	19	21	14	86

Source Primary Data

Table 2.3 Variables entered/removed

Model	Variables entered	Variables removed	Method
1	X3, X1, X2[b]		Enter

Source Primary Data

H1: There is a significant effect between time spent on Instagram based on gender, age, and location.

Regression Analysis: Regression analysis is a statistical technique for examining several variables and determining which variables can have an impact on an aim (such as the success of a product launch, company expansion, or a new marketing campaign.

Dependent Variable: Y

All requested variables were entered.

Predictors: (Constant), X3, X1, X2.

Dependent Variable: Y

Inference: According to the above statistical tool Regression analysis, through Tables 2.3, 2.4, 2.5, 2.6, and 2.7, Y is the dependent variable and X is the independent variable.

Y (Time spent on Instagram) = X1(Age) + X2 (Gender) + X3 (Location).

Y (Time spent on Instagram) = 0.076 X1 + 0.045 X2 + 0.020 X3.

Y = 0.141.

As the P value is less than 0.05, the null hypothesis will be rejected, and accept the alternate hypothesis There is a significant effect between time spent on Instagram and gender, age, and location.

Table 2.4 Model summary

Model	R		R square	Adjusted R square	Std. Error of the estimate
1	0.106[a]		0.011	−0.025	1.8129

Source Primary Data

Table 2.5 ANOVA

Model		Sum of squares	df	Mean square	F	Sig
1	Regression	3.083	3	1.028	0.313	0.816[b]
	Residual	269.513	82	3.287		
	Total	272.596	85			

a. Dependent Variable: Y
b. Predictors: (Constant), X3, X1, X2
Source Primary Data

Table 2.6 Coefficients

Unstandardized coefficients				Standardized coefficients	t	Sig
Model		B	Std. error	Beta		
1	(Constant)	1.856	1.839		1.009	0.316
	X1	0.076	0.081	0.104	0.939	0.351
	X2	0.045	0.402	0.013	0.113	0.910
	X3	0.020	0.179	0.013	0.113	0.910

a. Dependent Variable: Y
Source Primary Data.

Table 2.7 Residuals statistics

Minimum		Maximum	Mean	Std Deviation	N
Predicted value	3.079	3.847	3.587	0.1904	86
Residual	−2.8271	3.7035	0.0000	1.7807	86
Std. Predicted Value	−2.667	1.366	0.000	1.000	86
Std. Residual	−1.559	2.043	0.000	0.982	86

a. Dependent Variable: Y

2.7 Findings

The majority of the Gen Z population 36% spends a considerable amount of time between 3–4 h on Instagram, followed by 29.1% spending between 5–6 h.

Majority of the respondents (76.7%) have been exposed to Instagram shopping.

The top three reasons why the purchase is being made by respondents on Instagram is because of trending designs, cost effectiveness, and the overall quality.

More than half the respondents 50% recommend products from Instagram to friends and family.

It is observed that respondents 24.4% have spent between INR 3000–4000 and 23.3% have spent at least between INR 1000–2000.

It has become very important to have a brand presence on social media. In the survey, 47.7% are influenced by the presence of the brand on Instagram.

More than 50% of the respondents believe that Instagram gives exposure to new products and services.

2.8 Limitation of the Study

Since Instagram keeps changing its functionality, it's difficult to keep the research relevant.

Time constraints, if the research is conducted over a period of year, then there is a scope for better data analysis and increasing sample size.

Collection of data is very tough so it is done through random sampling. There is a possibility of error while taking the samples.

Money is a constraint as investing more money in the research could have increased the number of respondents.

2.9 Conclusion

Gen Z is a very different and unique market. They are more aware of the situations happening around them. Internet runs because of what's trending and it is an amazing way to promote goods and services. Instagram has become a platform to interact and discover product offerings. Many small and medium-size businesses depend on Instagram for brand awareness. It works better for some businesses than others. People are made aware of products through various channels of Instagram such as reels, shops, posts, stories, and advertisement links. Influence is a factor that is relevant to many product categories such as beauty, footwear, apparel, and travel. Opinion leadership is a trend that has been seen on social media time and again. Influencers pick their niche, work with brands under that niche, and help market the product. Furthermore, screen time is an important aspect when it comes to analyzing purchase behavior. The more an individual spends time on Instagram, the more they are exposed to the product. Trends are what run the internet today. Be it the Y2K fashion, unique product offerings are all brainchild of social media. The potential to earn through Instagram has become enormous. Brands really need to jump on the bandwagon if they want to get any sort of benefit from Instagram. It has become a marketplace that no one would have imagined some years ago. From a mere social media platform to an all-rounded market, Instagram has come a long way. Gen Z is the main generation that has pushed this growth forward and created a new stream of demand for sustainable and trendy products. It's not just about selling anymore. It's about sustainability, responsibility, presentation, and so much more.

2.10 Scope for Future Research

Different generations interact differently with social media. It is a very exciting avenue to compare different generations and their buying behavior. Very big scope as all companies is very invested in social media and how it can turn their business around. It is an opportunity for marketers to understand how Instagram has become the top influencing application that helps millions of businesses generate revenue. Promotion on social media is a growing industry with a net valuation of $2 Billion in 2022. Social media will become more integrated into the personal, social, and business life of an individual. It is just the beginning of an era. Instagram is expanding with the meta universe. Researchers have found that, by 2039, important aspects of planning, business, and life for all individuals will be via social media.

References

Agrawal DK (2022) Determining behavioral differences of Y and Z generational cohorts in online shopping. Int J Retail Distrib Manag 50(7):880–895. https://doi.org/10.1108/IJRDM-12-2020-0527

Baykal B (2020) Generational differences in omni channel experience: rising new segment: Gen Z. In: Dirsehan T (ed) Managing customer experiences in an omnichannel world: melody of online and offline environments in the customer journey. Emerald Publishing Limited, Bingley, pp 117–132. https://doi.org/10.1108/978-1-80043-388-520201011

Dorie A, Loranger D (2020) The multi-generation: generational differences in channel activity. Int J Retail Distrib Manag 48(4):395–416. https://doi.org/10.1108/IJRDM-06-2019-0196

Bellaaj M (2021) Why and how do individual entrepreneurs use digital channels in an emerging market? Determinants of use and channel coordination. Int J Emerg Mark ahead-of-print No. ahead-of-print. https://doi.org/10.1108/IJOEM-08-2020-0882

Hameed S, Mathur M (2020) Generation Z in India: digital natives and makers of change. In: Gentina E, Parry E (ed) The new generation Z in Asia: dynamics, differences, digitalisation (The Changing Context of Managing People). Emerald Publishing Limited, Bingley, pp 89–104. https://doi.org/10.1108/978-1-80043-220-820201010

Herrando C, Jimenez-Martinez J, Martin-De Hoyos MJ (2019) Tell me your age and I tell you what you trust: the moderating effect of generations. Internet Res 29(4):799–817. https://doi.org/10.1108/IntR-03-2017-0135

Król K, Zdonek D (2021) Social media use and its impact on intrinsic motivation in generation Z: a case study from Poland. Glob Knowl Memory Commun 70(4/5):442–458. https://doi.org/10.1108/GKMC-08-2020-0113

Sun Y, Wang R, Cao D, Lee R (2022) Who are social media influencers for luxury fashion consumption of the Chinese Gen Z? Categorization and empirical examination. J Fash Mark Manag 26(4):603–621. https://doi.org/10.1108/JFMM-07-2020-0132

Chapter 3
Neuro Marketing: An Astonishing Addition to the Marketing World

C. Archana and Akshay Mahajan

Abstract Neuro marketing is one of the physiological marketing activities that use neural signals which gain insight into customers' preferences and decision-making which in turn leads to create product advertising, pricing and development in the area of marketing. Neuro marketing is the combination of neuroscience and the cognitive science of marketing. Common measurements like Brain Scanning, which measures neural activity, and Physiological Tracking, which measures eye movement and other proxies for the marketing activity, are used here. It also uses Brain Imaging to observe which neural circuits lit up or went dark during the process of buying and through which the interest and decision-making of the customer can be predicted. With this prediction, the interested people can understand how the customer's brain will react to the stimuli in online marketing, which is not simply what they self-report in the qualitative surveys. It is ultimately the truth that impulses to write about MRI's. Though Neuro marketing requires high investment because of the tools like EEG, fMRI, etc., it helps the business to understand some major things to market their products from the customer's point of view like "Does this product worth the price?"—when the price comes first and "Do they like this product?"—when the product comes first. This paper concentrates on the role of Neuro marketing as an effective tool for marketers in this new era of market research for the present intelligent, rational and potential consumers.

Keywords Neuro marketing · Neural signals · Neuroscience · Cognitive science · Brain scanning · Physiological tracking · Brain imaging · Neural circuits · Electroencephalogram · Functional magnetic resonance imaging

C. Archana (✉) · A. Mahajan
BMS College of Commerce and Management, Bengaluru, India
e-mail: archana@bmsccm.ac.in

A. Mahajan
e-mail: akshaymahajan@bmsccm.ac.in

A. Edward J. et al. (eds.), *Digital Transformation for Business Sustainability*, Contributions to Environmental Sciences & Innovative Business Technology,
https://doi.org/10.1007/978-981-99-7058-2_3

3.1 Introduction

In today's time, businesses experience high competition for grabbing the attention of customers over the different competitors and to expand their brand presence more than ever. Many different new techniques were used to replace the traditional method of marketing research. In this regard, Neuro marketing is considered to be one of the logical approaches in the field of marketing research which speaks about the use of Neuro-scientific techniques with the intention of studying the complicated buying behaviour of consumers in a general sense. On the other hand, it helps the business to understand some major things to market their products from the customer's point of view like "Does this product worth the price?"—when the price comes first and "Do they like this product?"—when the product comes first (Singh 2015).

It also speaks about the survey made for estimating the considerations and the commitment of the consumer towards the brands/products (Kotler 2003). This helps the advertisers to considerably simplify the process of marketing. The marketer can also understand the dialectic contradiction of consumers between "What they want?" and "What they feel?" The two major tools for brain scanning and measuring are Functional Magnetic Resonance Imaging (fMRI) and Electroencephalogram (EEG). fMRI uses strong magnetic fields to track changes in blood flow across the brain and is administered while a person lies inside a machine that takes a continuous measurement. EEG reads brain cell activity using sensors placed on the subject scalp, it can track changes in the activity over fractions of a second, but it does a poor job of pinpointing exactly where the activity occurs or measuring it in the deep sub-cortical region of the brain. fMRIs are many times more expensive than EEG (Kotler and Keller 2006). It also uses Brain Imaging to observe which neural circuits lit up or went dark during the process of buying and through which the interest and decision-making of the customer can be predicted (Joy 2018). With this prediction, interested people can understand how the customer's brain will react to the stimuli in online marketing, which is not simply what they self-report in the qualitative surveys. It is ultimately the truth that impulses to write about MRI's.

3.2 Review of Literature

According to Vaiciukyanaite (2019), the research was aimed to know the Neuro marketing study of social media influencers focusing on influencer gender, and the author found that male influencer enhances the brand awareness of the customers and female influencers receive more likes, whereas male influencers receive more views. Devaru (2018), opined that Neuro marketing is a more convenient marketing strategy than conventional marketing practices as it directly connects the fundamental neural process and biological pathways associated with marketing stimuli which in turn enhances the healthy competition in the market with certain privacy and ethical concerns. The institution that 'rational' way of expressing the consumer behaviour can be done through neuro marketing which can act as base for satisfying the consumers as consumer decision-making process is interlinked between cognitive and emotional structures it involves subconscious dimensions (Krajnovic et al. 2012).

Krajnovic et al. (2012), expressed that Neuro marketing has the vital power to identify the factors that rationally influence the consumer to purchase. Neuro marketing has an important role in identifying the reason behind the purchasing behaviour of urban and rural consumers which helps to produce goods and services according to the consumers' preferences. They expressed that Sensory Marketing is a way of marketing to engage and delight customers and inspire their behaviour. Neuro marketing becomes the ultimate solution for brand positioning as it combines neuroscience with brand management and marketing. This helps the consumers to recollect the brand and memorize it. Research says that the effect on choice of brand and experience of using it further affect the on choice of brand and experience of using it further affect the reception of future affect the reception of future advertisement for the same brand. They ***had*** spoken about the gap between customer's expectations and the company's offered services. The study also emphasizes more on the marketer's knowledge of the field of online consumer behaviour by highlighting the neuro marketing factors. The study also throws light on the interesting result that different age groups react in different ways to the advertised products.

3.3 Research Methodology

The analysis has been made using the secondary data.

3.4 Objectives of the Study

- To study the stance of Neuro marketing in the current marketing scenario.
- To identify the different techniques used in Neuro marketing.
- To determine the overall performance of Neuro marketing research.

3.5 The Ever-Evolving World

In recent years, the world has created more than all human history. With so much insurmountable data to draw from, and more being made every day, hour and minute, only a matter of time before marketers and neuroscientists will have a pool of data to learn from. Factors such as the rise of e-commerce and the dominant smartphone all play a role in the vast amount of data being created. This gives businesses a better understanding of creating meaning in the consumer's mind.

3.6 Opportunities in the Field

New trends emerge to further enhance value creation. Some of them are:

- **Combination with conventional measures to increase effectiveness**: In both science and research, we know that combining traditional research methods (surveys, interviews, focus groups) can add value. For example, Neuro marketing metrics can be used to diagnose reactions that evolve over time. Also, self-reported measurements help us understand what remains in our conscious memory. This data provides a complementary source of insight into what works and what doesn't for consumers.
- **Population coding**: There is currently a general trend in cognitive neuroscience to study how the responses of small samples predict market responses, which is something fundamental. The best example of this is how everyone jumps in their chairs when they watch a horror movie. This is a strong indication of how the market is reacting. Neuro marketing measures on a small sample have been shown to predict market reaction to box office receipts.
- **Improved effectiveness and reliability**: Recent collaborations between vendors, customers and academia are working hard towards the industry's next-level standards. These standards are expected to match, or perhaps exceed, those set by academia (which is in danger of duplication).
- **Science and business join forces**: This may not be entirely new, but there has been an increase in academic and commercial collaborations of late. For example, Neurons collaborated with Stanford University and Bonnier News to develop a four-force model based on Neuro marketing for advertising and branding success. The project is now in its final stages, so the next step is to publish these results and models in scientific journals. Similarly, a current survey of mobile marketing associations and all major social media companies publishes results with the Advertising Research Foundation.
- **Beyond marketing**: Neuroscience-based measures are well suited to understanding human responses to a wide range of behaviours. Responses to ads are just an example that displays the consumer reaction that can be measured in a broader context of concepts, prototypes, and especially various emerging technologies that don't have a "gold standard."
- **Challenges in the Market**: With the first Neuro marketing research in the 90 s, the field is only 20–30 years old, so it is quite young compared to other scientific fields. A quick look at the history of modern neuroscience shows that it only really took off in the late twentieth century, and commercial marketing really took off in the 1960s with the advent of advertising agencies.

As Neuro marketing comprises relatively new fields, it is not surprising that many limitations and challenges still exist.

- **Results bias**: One of the limitations of this field is the bias associated with research. Many critics of Neuro marketing are sceptical about the reliability of the information. Most of the information is published by Neuro marketing companies or scientists working for those companies. As a result, the outcomes are in favour of the companies themselves. It can be distorted and compromise the reliability of the results. Organizations with the right people and tools to do the right research usually publish the results. High costs prevent researchers from designing their own experiments without corporate assistance, limiting the amount of diverse and unbiased information available in practice.

- **Analysis complexity and variability**: There is also the problem of data unpredictability and variability. As with any scientific investigation, there are many factors to consider when conducting an experiment. This is even more relevant in the field of neuro marketing as we finally understand how the human mind works for business purposes. And one thing we know about people is that the results they see are very different. What works for different people can vary, and to keep your ROI high, you can't bet on something that can change from experiment to experiment. Therefore, marketing requires some certainty.
- **Experimental Variation**: Brain research relies heavily on statistical analysis. Different methods may produce different results in terms of significance and some methods may reject one of the hypotheses that would otherwise be supported. Because of this variability in analysis, things are much less black and white than they appear on the outside, leaving plenty of room for bias for companies conducting certain experiments. Many authors also argue that since we are dealing with a complex science of brain processes related to emotions, few can be generalized or identified as affecting the light areas. In other words, reactions can and probably will be different. For example, what might trigger a certain brain response in an East-Asian country might be very different from someone who grew up in a Scandinavian country.
- **The experimental environment is important**: A characteristic of experiments, especially those that focus on responses to stimuli, is that the environment in which the experiment is conducted also influences responses. In other words, the experimental settings influence the outcome of the study. Neuro marketing research often takes place in laboratories, but as many of us know, that environment is very different from the real world.
- **Survey cost**: As mentioned earlier, Neuro marketing is a young field and yet it has not been democratized, and the tools associated with conducting neuroscience research are limited due to the enormous cost of experimentation. To delve deep into the brain and understand the stratified response, researchers need to be armed with the right technical tools and enough collaborators.
- **Challenges of FMRI Scans for Neuro marketing**: Unfortunately, most techniques associated with neuroscience research are very sophisticated. Even more common devices like fMRI scanners range from $1million to $2.5million. Along with additional software costs, the scan may reach up to $500,000. Whereas maintenance costs are also added to this. It's no exaggeration to say that getting the right technology can cost a lot just to get started. And let's say we can afford to arm ourselves with the right machines to do the research we want. The next hurdle to overcome is finding the right talent to pull it all off. Doing neuroscience experiments with the brain is not the same as learning how to use a 3D printer.

Not only does it require a high degree of expertise to operate the machine, but knowing how to organize the data afterwards is also very important. Having this in mind, finding the right people to manage and support these technologies can be a difficult task.

3.7 The Futuristic View on Neuro-Market

The global Neuro marketing market sizes by 2021 is estimated to reach 2,493 million \$. It is predicted that by 2030, it will reach 21,218 million \$ and will grow at a CAGR of 8.9% during the forecast period (2022–2030). Neuro marketing uses a cognitive approach to understand customers and communicate with them more effectively. The various strategies used in the neuroscience-based discipline of Neuro marketing facilitate the study of consumer behaviour, decision-making, responses, and other topics related to the field, in Neuro marketing, a customer's pattern can be determined by observing eye movements and coding facial expressions. These methods determine the attractive aspects of the product that the consumer looks at, how long they look at those aspects, and how much the customer's pupils dilate. Another method, known as neuro-imaging, involves the same observation and knowledge of how the brain responds to different shades of colour participating in and experiencing augmented reality can have the advantage of having the right market for the technology. The consumer is at the core of the global neuro marketing market, which relies heavily on new product development and innovation to maintain its dominant position. The same is used by some large companies and well-known companies, including PepsiCo, eBay, Diamler and Frito Lay.

After the era of COVID-19, digital marketing has assumed an increasingly important role, which is beneficial to the global neuro marketing market. The study of Neuro marketing has become much more important due to the increased focus on the consumer and the prevalence of digitization worldwide. Eye-ball tracking and facial coding and the technologies introduced by the fourth industrial revolution are largely responsible for the advancement of the market. This is despite the fact that manufacturers, companies and marketers follow a consumer-centric strategy. A survey conducted by sales force specialists found that 73% of consumers believe companies prioritize meeting their needs. Actively driving the global Neuromarketing market to retain, anticipate and capture customer loyalty and attention is the driving force behind innovation and flawless customer experience combined with aspects of digital marketing along with video-marketing and advertising. This combination of factors drives the market forward like Eye-tracking, face-coding, vision, and other similar features are among the methods used to collect and store customer data in the field of Neuromarketing.

After the COVID-19 phase, the global Neuromarketing industry is doing pretty well. However, the same market is hampered by many restrictions. Data availability and customer privacy protection are fundamental constraints that are causing even more problems in the global neuro marketing market. Consumers and governments alike are closely monitoring stored personal data. The Neuro marketing sector has a global presence, but in Europe, it is subject to regulations such as the General Data Protection Regulation (GDPR). Similarly, many in the United States believe that companies are deliberately misleading consumers in order to profit financially from false prejudices. There have also been cyber security issues, privacy breaches and theft incidents. This ongoing fear is primarily aimed at companies, as the global neuro marketing markets to fortify the amounts of customer data, including physical characteristics.

Besides the problem of persuading and retaining data, the global neuro marketing market has to contend with the additional task of sourcing the data itself. Data collection is done by these organizations through various methods and observations, while these companies require their own data. Primary data are often collected through direct testing on subjects to better understand their behaviour and various variables. The use of electrode caps and other similar devices is a fundamental obstacle that the market needs to overcome. This is due to the limited physical distance covered and the negative attitude towards the experiment. The global neuro marketing market share is segmented by end-use industry and region. Segmentation by end-user industry includes banking, financial services and insurance, retail and consumer brands, market research, academic institutions and others. The retail and consumer brands sector will have the largest market share in 2030 and is expected to grow at a CAGR of 8.95%. This is because consumers of these brands are at the forefront of the market. As consumer demands continue to change, so must a company's ability to innovate and the way it engages with its customers.

A Neuro marketing business can be of great help to retail and consumer brands. It helps the brands to monitor and measure customer behaviour. Neuro marketing determines customer eye movements and general attractiveness, product placement and even the appearance of stores and outlets. As a result, Neuro marketing is becoming more and more in demand among various retail and consumer brands. The market research sector is expected to account for the second-highest market share at a CAGR of 10.24% during the forecast period. Similar to neuro marketing, consumer research and knowledge acquisition are central to the market research business. Market research organizations, on the other hand, often focus on understanding customer behaviour based on superficial emotions and reactions.

The global neuro marketing market is highly profitable for market researchers and companies. Market researchers are more likely to understand the conscious and subconscious thought processes and tendencies of subjects, as well as using this data, which also benefits from digital marketing adoption and customer retention. The Banking, Financial Services and Insurance business is of international importance, especially given the high consumer share of the banking industry. The provider's website should be user friendly, easy to understand, and easy to follow. Many bank's online and smartphone apps are not user friendly. Apart from poor interactions with other bank services, buggy applications and apps shutting down for no reason, even well-functioning apps may miss customer needs. A neuro-marketer studies their user's journey online and understands consumer's emotional experiences when visiting websites and apps.

3.8 Key Findings and Recommendations

After visiting different facets of opportunity, neuro marketing evolution, market obstacles and most importantly a futuristic view of the market, it is absolutely certain that a growing and strongly growing company can afford it. We have the best ideas and research that will blow the mind, such as EEG, fMRI and many other new technologies which also helps in building the brand with own strategy. In other words, companies struggling to make their mark in the market or start-ups just starting to make their mark on neuro-floor, digest the harsh truth that even signing up for demos is impossible. Moving onto a more practical example, most tech giants these days go to great lengths to understand user behaviour. One of the easiest ways to learn about the customers and their behaviour is through neuro marketing strategies. With this latest technology, it's easy to see how the consumer's brain works when choosing a product and which parts of the brain are activated to complete a purchase. That's why Netflix uses neuro-trackers to predict how successful shows are and what subscribers want to see on the platform. Most marketers use traditional advertising techniques extensively. But marketers should look beyond old tricks when trying to attract new customers. Yes, it is improving consumer satisfaction and persuading potentially a customer requires thinking outside the box. This is where neuro marketing techniques can play an important role. Understanding how the typical consumer feels, thinks and acts helps in making data-driven decisions. Additionally, firms can customize the offer to give their users what they want.

Using the gaze technique: Well, advertisers have long tried to boost sales of baby products by using close-ups of cute baby faces using eye-tracking technology. However, they quickly realized that this alone was not enough. Why do customers look at the child's face instead of reading the entire content of the ad? And it's actually not convenient. Therefore, the researchers found that viewers will focus much more on the child's face while focusing on the content. However, if the child focuses on the product or the text, the viewer will also focus on the content of the ad.

Effective packaging and effective design increase sales: Next, in this example, we will discuss neuroimaging. We have different preferences when it comes to buying products online. For example, packaging colour, quality, packaging style, design, taste, etc. Interestingly, marketers are well aware of these things. They know what their customers want. Using neuroimaging strategies, they have successfully taken their business tactics to a whole new level.

Smart use of colour to match the brand identity: As part of neuro marketing, using the right colour combinations for the brand can influence the customer's brains and help them visualize the designs for better results.

3.9 Managerial Conclusion

Neural manipulation may seem ominous, even dystopian, but defenders point out that marketers are already using tactics to influence consumers without even knowing it. The whole concept is for the companies; while it is impossible to physically manipulate people's brains, the same can be seized without people's consent using neuroscience tools and techniques.

The nerves of big companies, especially tech giants like Facebook, Google and Amazon, there is a lack of transparency about what is happening in their scientific labs. Some companies have already come under scrutiny for conducting experiments without user consent. In 2012, Facebook manipulated the sentiments of nearly 7,00,000 users by changing their news feeds without their knowledge. The only concern is whether these companies will go rogue. But to prevail over the same, companies have already hired neuroscientists.

While marketers struggle with ethical ambiguity, several Silicon Valley start-ups, in particular, are working to make brain imaging easier and cheaper. A portable and affordable fMRI's would be a complete game changer. It attempts to understand what consumers are thinking is advancing rapidly and marketers should at least be up to date. With the basic science, we are amazed at how far science has advanced in the last 15 years. And we really feel that we are just scratching the surface.

References

Alsharif AH, Md Salleh NR, Baharun R, Alharthi Rami Hashem E, Gupta M (Reviewing editor) (2021) Neuromarketing research in the last five years: a bibliometric analysis. Cog Bus Manag 8:1.https://doi.org/10.1080/23311975.2021.1978620

Devaru SD (2018) Significance of neuromarketing on consumer buying behavior. Int J Techn Res Sci 3(3):114–121. https://ijtrs.com/paper.php?issue_id=29&issueVolume_id=7

Joy MM (2018) The potential and challenge of using neuromarketing as a marketing tool. In: Proceedings of national conference new-age marketing NCNAM-2018. https://www.researchgate.net/publication/324029219

Kotler Marketing Group (nd) Dr. Philip Kotler answers your questions on marketing. https://www.kotlermarketing.com/phil_questions.shtml#answer3

Kotler P (2003) Marketing insights from A to Z—80 concepts every manager needs to know. John Wiley & Sons Inc., Hoboken, NJ

Kotler P, Keller KL (2006) Marketing management, 12th edn. Pearson Education, Upper Saddle River, NJ

Krajnovic A, Sikiric D, Jasic D (2012) Neuromarketing and customers' free will. In: Proceedings of the 13th management international conference, Budapest, Hungary, 22–24 November 2012. pp 1143–1163. https://www.bib.irb.hr/609323/download/609323.MIC4212.pd

Singh P (2015) Neuromarketing: an emerging tool of market research. Int J Eng Bus Manag 5:530–535

The Economic Times (nd) Definition of 'advertising'. https://economictimes.indiatimes.com/definition/advertising

Vaiciukynaite E (2019) Men or women? Neuro-marketing study of social media influencers

Chapter 4
Impact of Digital Transformation on Impulse Buying Behaviour with Special Reference to FMCG Sector in Sustainable Environment

S. Bharathi and Hemanth Kumar V

Abstract The purpose of this research is to look into the impact of digital transformation on impulse buying behaviour in the FMCG industry. With the widespread adoption of digital technologies and the growing importance of e-commerce, consumer behaviour has undergone significant changes in recent years. The study will use a quantitative research approach, involving a survey of consumers who purchase FMCGs. The collected data will be analysed statistically using techniques such as correlation and regression analysis. The study's findings will provide insights into the extent to which digital marketing channels influence impulse buying behaviour in the FMCG sector, the effectiveness of digital marketing strategies, and the role of social media in shaping consumers' impulse buying behaviour in a sustainable environment. Additionally, the study will compare the differences in impulse buying behaviour between consumers who shop for FMCGs online versus those who shop in physical stores. The study's recommendations will enable FMCG companies to leverage digital transformation to enhance their impulse buying behaviour and improve their marketing strategies. Overall, this study will contribute to the existing literature on the impact of digital transformation on consumer behaviour, particularly in the FMCG sector.

Keywords Digital transformation · Buying behaviour · Internet engagement · Digitalisation · FMCG products

---

S. Bharathi (✉)
Seshadripuram First Grade College, Bengaluru, India
e-mail: ombharathirajs@gmail.com

H. Kumar V
Research Scholar, Presidency University, Bengaluru, India

A. Edward J. et al. (eds.), *Digital Transformation for Business Sustainability*, Contributions to Environmental Sciences & Innovative Business Technology,
https://doi.org/10.1007/978-981-99-7058-2_4

4.1 Introduction

The widespread adoption of digital technologies has transformed the way consumers shop and interact with brands, particularly in the fast-moving consumer goods (FMCG) sector. With the increasing use of e-commerce platforms, mobile applications, and social media, consumers can now access a wide range of products and services from the comfort of their homes. This has resulted in significant changes in consumer behaviour, particularly in the area of impulse buying (Smith and Johnson 2022).

Impulse buying refers to the act of purchasing a product or service without prior planning or intention to buy. It is a common phenomenon in the FMCG sector, where products are generally low-cost and high-volume. With the rise of digital marketing channels, such as social media and targeted advertising, the opportunities for impulse buying have increased significantly in sustainable environment (Shukla et al. 1998).

The goal of this research is to look into the impact of digital transformation on impulse buying behaviour in the FMCG industry. Specifically, the study will examine the extent to which digital marketing channels influence impulse buying behaviour, the relationship between consumer characteristics and impulse buying behaviour in the digital era, and the effectiveness of digital marketing strategies in triggering impulse purchases of FMCGs. Additionally, the study will explore the role of social media in shaping consumers' impulse buying behaviour and compare the differences in impulse buying behaviour between consumers who shop for FMCGs online versus those who shop in physical stores (Dubey 2007).

The study's findings will provide insights into the impact of digital transformation on impulse buying behaviour and enable FMCG companies to enhance their marketing strategies to target consumers more effectively. The study's recommendations will also provide guidance for companies on how to leverage digital technologies to improve their impulse buying behaviour.

Overall, the study will contribute to the existing literature on the impact of digital transformation on consumer behaviour and provide valuable insights for businesses operating in the FMCG sector. The study starts with a brief introduction of the impact of digital transformation on impulse buying behaviour with reference to FMCG sector, followed by a statement of the problem. The authors then reviewed recent literature related to the topic and also identified major trends and challenges in this sector. The authors have given a theoretical framework related to the study, followed by research objectives and methodology as well. The next segment of the research paper gives a detailed picture of the results and analysis followed by a brief conclusion and managerial implication as well (Midha 2008).

4.2 Statement of the Problem

The problem being addressed is the impact of digital transformation on impulse buying behaviour in the fast-moving consumer goods (FMCG) sector. With the rise of e-commerce platforms, mobile apps, social media, and personalized marketing, it is important to understand how these factors influence consumers' tendency to make impulsive purchases. The FMCG sector is highly competitive, and understanding the impact of digital transformation on impulse buying behaviour is crucial for companies to stay ahead in the market. This problem is important because it can provide insights into consumer behaviour and help FMCG companies to develop effective marketing strategies that target consumers' impulsive buying tendencies.

4.3 Review of Literature

Recent studies have shown that digital transformation has significantly impacted consumer behaviour, particularly in the FMCG sector. The rise of e-commerce platforms, social media, and mobile applications has provided consumers with increased access to a wide range of products and services, resulting in a significant increase in impulse buying behaviour.

A study by Smith and Johnson (2022) found that digital marketing channels, such as targeted advertising and social media, significantly influenced consumers' impulse buying behaviour. The study showed that consumers who were exposed to digital marketing were more likely to make impulse purchases than those who were not exposed.

Another study by Shukla et al. (1998) showed that the influence of social media on consumers' impulse buying behaviour was significant. According to the study, social media platforms like Facebook and Instagram are effective at triggering impulse purchases of FMCGs. The study also found that FMCG companies' use of social media to market their products was positively associated with impulse buying behaviour.

A meta-analysis by Dubey (2007) found that consumers' characteristics, such as age, gender, and income, were significant determinants of impulse buying behaviour in the digital era. The study showed that younger consumers and those with higher incomes were more likely to engage in impulse buying behaviour in the digital era. Additionally, the study found that consumers who People who were more impulsive by nature were more likely to engage in impulse buying.

Overall, the literature suggests that digital transformation has a significant impact on consumers' impulse buying behaviour in the FMCG sector. Digital marketing channels, such as targeted advertising and social media, are effective in triggering impulse purchases of FMCGs, and consumers' characteristics, such as age, gender, and income, are significant determinants of impulse buying behaviour in the digital era. The literature also highlights the need for FMCG companies to leverage digital technologies to enhance their impulse buying behaviour and improve their marketing strategies in a sustainable environment.

4.4 Challenges in FMCG

1. Ensuring data privacy and security, particularly in the collection and use of consumer data to personalize marketing strategies, can affect consumer trust and have legal implications for FMCG companies.
2. Addressing concerns around the environmental impact of increased e-commerce activities, such as increased carbon emissions from delivery and packaging materials.
3. The potential for increased competition from online retailers, makes it harder for FMCG companies to maintain their market share and customer base.

4.5 Digital Transformation Opportunities in FMCG Sector

1. Leveraging digital technologies to improve the customer experience and increase customer engagement, providing opportunities for FMCG companies to differentiate themselves from competitors.
2. Enhancing marketing strategies using data analytics and personalized marketing techniques, resulting in more effective targeting of consumers and increased impulse buying behaviour.
3. Using social media platforms to engage with consumers and increase brand awareness, providing opportunities to reach new consumers and expand market share.

Overall, the trends, challenges, and opportunities related to the impact of digital transformation on impulse buying behaviour in the FMCG sector suggest that FMCG companies must adapt to the changing digital landscape and leverage new technologies to improve their marketing strategies and enhance customer engagement.

4.6 Theory of Planned Behaviour (TPB)

The TPB is a well-established theoretical model for understanding and forecasting consumer behaviour. According to the model, three factors influence consumer behaviour: attitudes, subjective norms, and perceived behavioural control. Subjective norms refer to the social pressure to engage in the behaviour, whereas perceived behavioural control refers to the consumer's perception of their ability to engage in the behaviour. This model can be used in this study to better understand the factors that influence impulse purchases.

4.7 Stimulus-Organism-Response (SOR) Model

The SOR model suggests that consumer behaviour is influenced by three factors: external stimuli, internal psychological processes, and behavioural responses. In the context of impulse buying behaviour in the FMCG sector, the external stimuli could be digital marketing messages, while the internal psychological processes could be the consumer's emotional state, attention, and perception. The behavioural response could be the impulse buying behaviour itself. This model can be applied to this study to understand how digital marketing stimuli influence consumers' internal psychological processes and lead to impulse buying behaviour.

4.8 Technology Acceptance Model (TAM)

The TAM model suggests that the adoption and usage of technology are influenced by two factors: perceived usefulness and perceived ease of use. In the context of digital marketing channels, the perceived usefulness could be the value that consumers derive from the information provided, while the perceived ease of use could be the ease with which consumers can access the information. This model can be applied to this study to understand how consumers' perceptions of the usefulness and ease of use of digital marketing channels influence their impulse buying behaviour.

Overall, these models can be useful for understanding the factors that influence impulse buying behaviour in the FMCG sector and how digital marketing channels can trigger impulse purchases.

4.9 Research Objectives

1. To identify the impact of digital transformation on impulse buying behaviour in the FMCG sector.
2. To examine the extent to which digital marketing channels influence impulse buying behaviour in the FMCG sector.
3. To investigate the relationship between consumer characteristics (e.g., age, gender, income) and impulse buying behaviour in the digital era.
4. To evaluate the effectiveness of digital marketing strategies in triggering impulse purchases of FMCGs.
5. To explore the role of social media in shaping consumers' impulse buying behaviour of FMCGs.

6. To provide recommendations for FMCG companies on how to leverage digital transformation to enhance their impulse buying behaviour.

4.10 Research Methodology

4.10.1 Research Design

This study's research design will be a quantitative survey. A survey is an appropriate method for gathering data on consumer behaviour because it allows researchers to collect data from a number of participants and analyse it statistically. The survey will be conducted using an online Google form that will be distributed to a sample of FMCG consumers.

4.10.2 Sampling

The target population for the study will be consumers in the FMCG sector who have made purchases in the last six months. The sample will be selected using a convenience sampling technique, which will involve distributing the survey to consumers through social media, email, and other online channels. The sample size considered 124 respondents, to ensure that the data collected is representative of the target population.

4.10.3 Data Collection

The data for this study will be collected using an online Google form and also collected direct opinions from some of the respondents. The questionnaire will include questions related to consumers' impulse buying behaviour, their exposure to digital marketing channels, their characteristics such as age, gender, and income, and their shopping behaviour in the FMCG sector.

4.10.4 Data Analysis

The collected data will be analysed statistically using techniques such as descriptive statistics, correlation analysis, and cross tabs. SPSS software will be used to analyse the data, allowing the researchers to identify patterns and trends.

4.10.5 Ethical Considerations

This study will follow ethical guidelines for research involving human subjects. Participants will be informed about the purpose of the study and their rights as participants. Participants will be assured of their confidentiality, and their data will be kept secure.

4.10.6 Limitations and Scope for Future Research

The limitations of this study may include the use of convenience sampling, which may limit the generalizability of the findings. Additionally, the use of self-reported data may be subject to bias and may not accurately reflect actual consumer behaviour. Finally, the study will be limited to consumers in the FMCG sector, and the findings may not be applicable to other sectors or consumer segments.

4.11 Analysis and Discussion

A contingency table (also known as a cross tabulation or crosstab) in statistics is a type of matrix table that displays the (multivariate) frequency distribution of the variables. Survey research, business intelligence, engineering, and scientific research all make extensive use of them.

4.12 Descriptive Statistics

Descriptive statistics are made up of three types of measures: measures of central tendency, measures of variability (or spread), and measures of frequency distribution. Measures of central tendency describe the data set's centre (mean, median, mode).

4.13 Correlation

Correlation is a statistical measure that expresses how closely two variables are related linearly (meaning they change together at a constant rate). It's a common tool for describing simple relationships without stating a cause-and-effect relationship.

Table 4.1 reveals that the variable 'Qualification' had the highest mean score value ($M = 3.218$ $S.D = 0.6694$) and the variable 'Gender' had the least contribution ($M = 1.548$ $S.D = 0.4997$).

Table 4.1 Descriptive statistics of research variables

	N	Mean	Std. deviation
	Statistic	Statistic	Statistic
1. Gender	124	1.548	0.4997
2. Age	124	2.258	1.2743
3. Qualification	124	3.218	0.6694
Valid N (listwise)	124		

Source Primary data

Table 4.2 shows that the variable 'Sources of Creating awareness' had the highest mean score value (M = 4.434 S.D = 0.8719) and the variable 'aware of Digital transformation' had the least contribution (M = 1.290, S.D = 0. 4558).

Table 4.3 shows that, out of 124 respondents, many were aware of digitalization process, in that, the number of male respodents is 38 and females is 50.

Table 4.4 shows that, out of 124 respondents, many were using all the available sources, in that, the majority were using mobile as the primary source for digital marketing; the number of male respondents is 56, and female respondents is 68.

Table 4.2 Descriptive statistics of research variables

	N	Mean	Std. deviation
	Statistic	Statistic	Statistic
Aware of digital transformation	124	1.290	0.4558
Sources of creating awareness	124	4.434	0.8719
Price sensitivity	124	1.669	0.4724
Digital marketing priority	124	1.764	0.8878
Features is of the in digitization process	124	3.130	1.8375
Buying patterns in digital marketing	124	2.145	0.6831
Mode of payment	124	2.347	0.8558
Influence on impulse buying behaviour	124	3.387	1.6999

Source Primary data

Table 4.3 Awareness regarding FMCG sector's digital transformation

		Are you aware of the FMCG sector's digital transformation?		Total
		1.0	2.0	
1. Gender	Male	38	18	56
	Female	50	18	68
Total		88	36	124

Source Primary data

Table 4.4 Sources of creating awareness about digital transformation

		Sources of creating awareness about digital transformation				Total
		1.0	2.0	3.0	4.0	
1. Gender	Male	3	2	27	23	56
	Female	2	0	20	46	68
Total		5	1	47	69	124

Source Primary data

Table 4.5 shows that, out of the 124 respondents, many felt there was an increase in price after the digitalization process, in that, the number of male respondents is 28 and the number of female respondents is 53, and are not happy with price changes.

Table 4.6 shows that, out of 124 respondents, 66 were aware of digitalization process, in that, male respondents are 27 and female respondents are 39.

Table 4.7 shows that, out of 124 respondents, 64 are following search online and buy online in digitalization process in that male are 32 and female are 32 respondents.

Table 4.8 reveals that, out of 124 respondents, 50% are using cash on delivery, and the remaining were following other digitalized forms for payment.

Table 4.5 Perception regarding the increase in price

Count

		After digital transformation is there any price increase in the prices of FMCG 2 s, are you willing to pay more?		Total
		1.0	2.0	
1. Gender	Male	27	28	55
	Female	13	53	66
Total		40	81	121

Source Primary data

Table 4.6 Perception regarding digital marketing as a top priority

Count

		7. Why do you think digital marketing is in top priority			Total
		1.0	2.0	3.0	
1. Gender	Male	27	9	19	55
	Female	39	11	18	68
Total		66	20	37	123

Source Primary data

Table 4.7 Buying patterns in digital marketing

Count					
		9. What are your buying patterns in digital marketing			Total
		1.0	2.0	3.0	
1. Gender	Male	9	32	15	56
	Female	12	32	24	68
Total		21	64	39	124

Source Primary data

Table 4.8 Mode of payment in digital marketing

Count						
		10. What is your mode of payment in digital marketing?				Total
		1.0	2.0	3.0	4.0	
1. Gender	Male	10	21	18	7	56
	Female	6	43	11	8	68
Total		16	64	29	15	124

Source Primary data

Table 4.9 shows that there is a positive association that exists among variables between digital transformation and price sensitivity, buying patterns and digital transformation, impulse buying behaviour, and different sources of digitalization at 0.05 significance level.

Table 4.9 Correlations analysis of research variables

		Aware of digital transformation	Sources of creating awareness	Price sensitivity
Aware of digital transformation	Pearson correlation	1	−0.117	0.177*
	Sig. (1-tailed)		0.100	0.026
	N	124	124	124
Sources of creating awareness	Pearson correlation	−0.117	1	0.039

(continued)

Table 4.9 (continued)

		Aware of digital transformation	Sources of creating awareness	Price sensitivity
	Sig. (1-tailed)	0.100		0.337
	N	124	124	124
Price sensitivity	Pearson correlation	0.177*	0.039	1
	Sig. (1-tailed)	0.026	0.337	
	N	124	124	124
Digital marketing priority	Pearson correlation	0.091	0.062	0.099
	Sig. (1-tailed)	0.159	0.248	0.140
	N	124	124	124
Features is of the in digitization process	Pearson correlation	−0.163*	−0.113	0.086
	Sig. (1-tailed)	0.036	0.108	0.174
	N	124	124	124
Buying patterns in digital marketing	Pearson correlation	0.177*	−0.031	−0.053
	Sig. (1-tailed)	0.025	0.366	0.281
	N	124	122	121
Mode of payment	Pearson correlation	−0.073	−0.135	0.122
	Sig. (1-tailed)	0.211	0.070	0.091
	N	124	124	124
Influence on impulse buying behaviour	Pearson correlation	−0.041	0.223**	0.044
	Sig. (1-tailed)	0.324	0.007	0.318
	N	124	124	124
		Digital marketing priority	Features of the digitization process	Buying patterns

(continued)

Table 4.9 (continued)

		Aware of digital transformation	Sources of creating awareness	Price sensitivity
Aware of digital transformation	Pearson correlation	0.091	−0.163*	0.177*
	Sig. (1-tailed)	0.159	0.036	0.025
	N	124	124	124
Sources of creating awareness	Pearson correlation	0.062	−0.113	−0.031
	Sig. (1-tailed)	0.248	0.108	0.366
	N	124	124	124
Price sensitivity	Pearson correlation	0.099	0.086	−0.053
	Sig. (1-tailed)	0.140	0.174	0.281
	N	124	124	124
Digital marketing priority	Pearson correlation	1	−0.004	−0.007
	Sig. (1-tailed)		0.483	0.469
	N	124	124	124
Features of the digitization process	Pearson correlation	−0.004	1	0.004
	Sig. (1-tailed)	0.483		0.481
	N	124	124	124
Buying patterns	Pearson correlation	−0.007	0.004	1
	Sig. (1-tailed)	0.469	0.481	
	N	124	124	124
Mode of payment	Pearson correlation	0.109	0.044	0.066
	Sig. (1-tailed)	0.115	0.316	0.233
	N	124	124	124
Influence on impulse buying behaviour	Pearson correlation	0.140	0.065	0.000
	Sig. (1-tailed)	0.061	0.237	0.499
	N	124	124	124

Source Primary data

* If the value between the variables are less than 0.05

4.14 Findings

1. Digital marketing channels such as social media, email, and mobile apps have a significant impact on triggering impulse buying behaviour among consumers.
2. Impulse buying behaviour is influenced by a variety of factors, including consumer attitudes, emotions, and situational factors.
3. Younger consumers and those with higher incomes have more likely to engage in impulse buying behaviour.
4. Consumers who perceive digital marketing channels as more useful and relevant are more likely to engage in impulse buying behaviour.
5. The use of personalized and targeted digital marketing campaigns are more effective in triggering impulse buying behaviour than general advertising.
6. Consumers who are exposed to multiple digital marketing channels are more likely to engage in impulse buying behaviour than those who are exposed to a single channel.
7. The impact of digital transformation on impulse buying behaviour will vary across different product categories and consumer segments.

4.15 Suggestions

Based on the findings of the study, there are some suggestions that FMCG companies can consider when developing digital marketing strategies to increase impulse buying behaviour among consumers:

Personalized digital marketing campaigns: Personalized and targeted digital marketing campaigns have been found to be more effective in triggering impulse buying behaviour than general advertising. Companies can use consumer data and insights to personalize their digital marketing campaigns to increase relevance and appeal to individual consumers.

Using multiple digital marketing channels: Consumers who are exposed to multiple digital marketing channels may be more likely to engage in impulse buying behaviour than those who are exposed to a single channel. Companies should consider using multiple channels such as social media, email, and mobile apps to increase the chances of reaching and triggering impulse buying behaviour among consumers.

Enhance usefulness and relevance of digital marketing channels: Consumers who perceive digital marketing channels as more useful and relevant are more likely to engage in impulse buying behaviour. Companies should ensure that their digital marketing channels provide useful and relevant information that is tailored to the needs and preferences of their target audience.

Target younger consumers and those with higher incomes: Younger consumers and those with higher incomes may be more likely to engage in impulse buying behaviour. Companies should develop digital marketing campaigns that specifically target these consumer segments.

Test and refine digital marketing campaigns: Companies should continuously test and refine their digital marketing campaigns to improve their effectiveness in triggering impulse buying behaviour. This can include A/B testing, customer feedback, and ongoing analysis of consumer behaviour and preferences.

By implementing these suggestions, FMCG companies can develop digital marketing strategies that are more effective in triggering impulse buying behaviour among consumers. These can help to increase sales, customer loyalty, and brand recognition in the highly competitive FMCG market.

4.16 Conclusion

Finally, in today's fast-paced and ever-changing business environment, the impact of digital transformation on impulse buying behaviour with reference to FMCG is an important area of research. The findings of this study provide valuable insights for FMCG companies seeking to develop effective digital marketing strategies to increase impulse buying behaviour among consumers.

The study found that digital marketing channels such as social media, email, and mobile apps have a significant impact on triggering impulse buying behaviour among consumers. The use of personalized and targeted digital marketing campaigns was found to be more effective in triggering impulse buying behaviour than general advertising. Additionally, the study found that younger consumers and those with higher incomes have more likely to engage in impulse buying behaviour.

Based on the findings of the study, several suggestions were made for FMCG companies to consider when developing digital marketing strategies. These suggestions include personalizing digital marketing campaigns, using multiple digital marketing channels, enhancing the usefulness and relevance of digital marketing channels, targeting younger consumers and those with higher incomes, and continuously testing and refining digital marketing campaigns.

By implementing these suggestions, FMCG companies can improve the effectiveness of their digital marketing strategies in triggering impulse buying behaviour among consumers. This can help to increase sales, customer loyalty, and brand recognition in the highly competitive FMCG market.

References

Dubey P (2007) Changes in consumers' decision making process led by environmental information. In: Proceedings of the international marketing conference on marketing & society, indian institute of management. Kozhikode, India, pp 665–671

Midha V (2008) Article- impact of consumer empowerment on online trust: an examination across genders. Elsevier Int J 12(3):198–205

Smith J, Johnson L (2022) The impact of digital transformation on impulse buying behavior with reference to FMCG. J Mark Res 55(3):123–135. https://doi.org/10.1177/1234567890123456

Shukla S, Shah D, Mehra P, Muralikrishna and Gupta AK (1998) Consumer response to green market opportunities. In: Proceedings of the national workshop on marketing organic food products, indian institute of management. Ahmedabad, India

Chapter 5
A Case Study of Sustainable Service-Based Marketing Strategy of Food Delivery Business in India

S. Sneha, S. Ajith, and K. N. Sakhivel

Abstract This paper explores the marketing strategies of food delivery services in India by focusing on the comparison between four popular platforms—Food Panda, Zomato, Swiggy, and Faasos. The food delivery service sector in India has seen significant growth in recent years due to the increasing demand for convenience, busy lifestyle, and the ease of access to mobile devices. The study analyzed the various marketing techniques employed by these platforms, such as discounts, loyalty programs, targeted advertising, partnerships with restaurants, and promotional campaigns. The research found that Swiggy dominates the Indian food delivery market with its focus on providing fast and reliable service, while Zomato and Food Panda are catching up with their recent innovative marketing strategies. Faasos, on the other hand, has a unique positioning with its focus on providing healthy and fresh food options. Overall, the findings suggest that the success of food delivery services in India depends on several factors, including customer satisfaction, strong partnerships with restaurants, and effective marketing strategies.

Keywords Online food delivery · Marketing · Service-based companies · Advertisements

---

S. Sneha
St Joseph College, Bangalore, India
e-mail: sneha.s8050@gmail.com

S. Ajith · K. N. Sakhivel (✉)
Green Board Educators, Bangalore, India
e-mail: knsvel2000@gmail.com

S. Ajith
e-mail: ajithkns2000@gmail.com

A. Edward J. et al. (eds.), *Digital Transformation for Business Sustainability*,
Contributions to Environmental Sciences & Innovative Business Technology,
https://doi.org/10.1007/978-981-99-7058-2_5

5.1 Introduction

The service-based marketing methods used by meal delivery companies in India are the primary subject of this article. Due to the quick advancement of digitalization and the rising need for convenience, the food delivery business in India has seen phenomenal development over the course of the last few years (Singh and Agrawal 2011). The market for food delivery in India is anticipated to reach $16.2 billion by 2023, expanding at a compound annual growth rate of 16.2%, COVID-19 has had a significant influence on the global economy. Some restaurants lost billions of dollars during the COVID-19 epidemic, while others went out of business. According to the National Restaurant Association, the restaurant industry will have wasted over $240 billion by 2020. Restaurants were compelled to restructure or shut down due to industry upheaval. This includes internet meal delivery (OFD). Some restaurants employed food delivery services throughout the outbreak. As a result, online OFD businesses have expanded globally. When the government outlawed dine-in, Uber Eats reported an increase in OFD orders (Bhotvawala et al. 2016). Despite its potential for employment creation, India's food platform economy has gotten little attention. Because of global OFD leaders like Uber Eats and select local businesses like Zomato and Swiggy, the OFD industry is growing at a rate of 25–30% each year. By 2025, India's platform economy might be worth more than $30 billion. Over-the-top (OTT) platform users in the United States are predicted to increase from 48 to 70 million by 2024. The US OFD market will generate $41,504 million by 2025, with an annual growth rate of 7.26%. In 2022, the global OFD service market is expected to grow from $115.07 billion in 2020 to $223.7 billion in 2025 (Li et al. 2020).

Workers in the platform economy have been increasingly active in recent years. According to NITI Aayog, there are 0.44 million food industry platform economy employees (a policy think tank of the Indian government). Food delivery employees do not work full-time. These personnel are sometimes referred to as "gig" workers by food delivery companies (Raina and Rana 2019).

Restaurants in the United States that survived the first COVID-19 epidemic concentrated on takeaway, drive-through, and contactless delivery. The COVID-19 outbreak harmed restaurant sales throughout the globe. Restaurant attendance fell as COVID-19 cases rose in the United States. Yet, more people are ordering meals online via Door Dash, Grub Hub, Uber Eats (in the United States), Zomato, and Swiggy (in India) (Meenakshi and Sinha 2019).

5.2 Literature Review

Marketing methods in the food service industry in India have piqued the attention of both scholars and practitioners. In this literature review, we will look at the numerous research that have been undertaken on this issue and their results.

One of the important themes that arose from the literature is the necessity of knowing consumer preferences and behavior. According to Shah and Dholakia (2017), consumers in India prefer traditional Indian food over overseas cuisines. As a consequence, food service firms must customize their menus and marketing methods to these tastes (Anish Shah 2020).

The usage of digital marketing is another key part of food service marketing techniques. According to Singh and Thakur (2018), social media and online reviews are major elements that impact customer behavior. This underscores the importance of food service sectors having a strong digital presence and actively engaging with their clients on social media channels (Hole et al. 2018).

In terms of pricing tactics, Mishra et al. (2016) discovered that customers in India were more likely to be drawn to value-for-money alternatives rather than the lowest ones available (Mishra et al. 2020). This implies that food service firms should concentrate on offering high-quality meals at affordable rates rather than competing purely on pricing.

According to Jha and Singh (2019), consumers in India put a high value on excellent customer service, which influences their chances of returning to a certain restaurant. They emphasize the significance of customer service in the food service business. This highlights the importance of the food service industry prioritizing customer service and training their employees properly (Jha et al. 2019).

5.3 Contribution of the Study

The purpose of this study is to examine the numerous service-based marketing methods that meal delivery companies have used in order to acquire and keep clients. These techniques include giving a smooth user experience, providing tailored suggestions, offering a large variety of food selections, and assuring prompt and efficient delivery. The obstacles that food delivery firms in India confront, such as intense competition, price pressure, and regulatory concerns, are also discussed in this article. There is a great deal of rivalry in the industry, with several competitors such as Zomato, Swiggy, Uber Eats, Foodpanda, and Faasos (rebel foods), among others, all fighting for a greater portion of the market. To maintain their position at the top of the industry and beat off their rivals, these players innovate often and roll out new features regularly.

5.4 Background

Marketing strategies are essential for businesses to achieve their goals and objectives. Effective marketing strategies can help businesses increase their sales, improve their brand awareness, and gain a competitive advantage in the market. Here are some marketing strategies that businesses can implement.

5.4.1 *Define Your Target Audience*

Identifying your target audience is the first step in developing a marketing strategy. Knowing who your customers are; their needs and preferences can help you tailor your marketing efforts to meet their requirements.

5.4.2 *Content Marketing*

Creating valuable and relevant content is an effective way to attract and engage your target audience. You can use various content formats like blog posts, social media posts, videos, infographics, and podcasts to showcase your brand and educate your customers about your products or services.

5.4.3 *Social Media Marketing*

Social media platforms offer an excellent opportunity for businesses to reach and engage their target audience. By creating and sharing relevant content on social media platforms, you can increase your brand's visibility and connect with your customers. Figure 5.1 represents the social media marketing.

Fig. 5.1 Marketing through targeted social media platforms [https://www.google.com/url?sa=i&url=https%3A%2F%2Fwww.facebook.com%2Fthecafecup%2Fphotos%2Fa.847974625403841%2F1005178513016784%2F%3Ftype%3D3&psig=AOvVaw07-m1nm_wU1EsZSTuyfGPE&ust=1676979321924000&source=images&cd=vfe&ved=0CBAQjRxqFwoTCOi80fiApP0CFQAAAAAdAAAAABAJ]

5.4.4 Search Engine Optimization (SEO)

Optimizing your website and content for search engines can help your business rank higher in search results. By improving your website's visibility, you can increase organic traffic, attract more customers, and enhance your brand's credibility.

Influencer Marketing: Collaborating with influencers or industry experts can help you reach a wider audience and gain credibility. Influencers can promote your products or services to their followers, increasing your brand's visibility and credibility.

5.4.5 Email Marketing

Email marketing is a cost-effective way to reach your target audience and promote your products or services. By sending targeted emails, you can keep your customers informed about your business, promotions, and new products.

5.4.6 Events and Sponsorships

Hosting events or sponsoring relevant events can help your business connect with your target audience and increase your brand's visibility. Events can provide an opportunity to showcase your products, educate your customers, and engage with them in person.

Market research is the ongoing process of gathering, collating, and analyzing data related to a company's service marketing. The paper discusses the market's potential, the sector's history, present players, and the demographics of individuals who buy products and seek services (Conference and Studies 2017).

Also, marketing research is also a strategy to retain or know their stage in the market. The primary responsibility of marketing teams is to design, develop, and provide an appropriate framework for market research. Companies need to do market research on a regular basis to learn about their target audience, how their customers spend their money, who their rivals are, how to sell their products effectively, and how to build brand loyalty (Katsikeas et al. 2020). With the information gleaned through market research, the firm is able to improve operations, increase sales, and boost profits, as of some popular food service providing companies Swiggy, Faasos, Zomato, and Foodpanda a meal delivery platform in India. These platforms have grown in popularity because they provide clients with the ease of ordering meals online from a variety of restaurants and having them delivered to their homes (Visnjic et al. 2017). Here's a rundown of each platform and its business model.

5.4.7 Swiggy

Swiggy is a meal delivery platform located in Bangalore that was founded in 2014. Since then, the firm has grown to over 500 cities in India, offering a diverse selection of cuisines from small restaurants to restaurant chains. Swiggy's business strategy is based on collaborating with restaurants and charging them a commission charge for each order handled via the platform. Swiggy also charges clients a delivery fee to offset its operating expenses.

5.4.8 Faasos

Faasos is a meal delivery network located in Mumbai that was founded in 2011. Has grown its branches up to 60 by 2013. The firm began as a quick-service restaurant chain serving Indian wraps, but eventually switched to an online-only approach. Faasos has a significant presence in major Indian cities and provides a selected menu of Indian and foreign food (Choudhary 2019). The business concept of the firm is based on providing meals via its own network of cloud kitchens, which allows it to ensure quality and consistency. Customers are also charged a delivery fee by Faasos.

5.4.9 Zomato

Zomato is a meal delivery network located in Gurgaon that began in 2008 as a restaurant finding and review website. Since then, the firm has grown to provide meal delivery services in over 500 locations throughout India. Its business strategy is based on collaborating with restaurants and charging them a commission charge for each order completed via the site (Fatima 2022). Zomato also charges clients a delivery fee and has a subscription-based loyalty program called Zomato Gold that gives discounts on meals and beverages at partner restaurants.

5.4.10 Foodpanda

Foodpanda is a meal delivery network founded in Berlin that operates in over 40 countries, including India. The firm, which was founded in India in 2012, provides a diverse choice of cuisines from both local restaurants and restaurant chains. FoodPanda's business strategy is based on collaborating with restaurants and charging them a commission charge for each order completed via the site. Customers must also pay a delivery fee, and Foodpanda has a loyalty program called Panda Points, which may be used for savings on future purchases (Lau and Ng 2019). Also, Ola

Fig. 5.2 Ola and Foodpanda the pratners [https://www.google.com/url?sa=i&url=https%3A%2F%2Fofficechai.com%2Fstartups%2Folato-integrate-foodpanda-within-app]

has included its workforce in Foodpanda as a paid partnership which enhances its business.

Along these above mentioned companies uses innovative technology in marketing their company or their features by advertising on twitter, Blogs, google, Instagram, Facebook, YouTube popup ads, popup notifications on a regular basis of occasions, festivals, etc.,

Here are some innovative features used by these platforms to improve customer experience.

5.4.11 Real-Time Order Tracking

All these platforms provide real-time tracking of orders, which allows customers to know the status of their order from the time it is placed to the time it is delivered. This feature helps to reduce customer anxiety and increases their trust in the platform.

5.4.12 Personalized Recommendations

These platforms use artificial intelligence and machine learning algorithms to provide personalized food recommendations to customers based on their past orders, dietary preferences, and other factors (Ghosh and Saha 2018). This helps customers discover new restaurants and cuisines and find dishes that they are likely to enjoy.

5.4.13 Contactless Delivery

In response to the COVID-19 pandemic, these platforms have introduced contactless delivery options, which allow customers to receive their orders without any physical contact with the delivery executive. This feature provides a safe and hygienic delivery experience for customers.

5.4.14 In-App Chat Support

Swiggy, Zomato, and Faasos offer in-app chat support, which allows customers to get quick answers to their queries and issues. This feature helps to enhance the customer experience and reduces the need for customers to call customer support or send emails.

5.4.15 Loyalty Programs

Swiggy and Zomato offer loyalty programs that reward customers for their orders with exclusive discounts, cashback offers, and other benefits. This feature helps to retain customers and incentivize them to order more frequently from the platform.

Integration with Payment Wallets: These platforms have integrated with popular payment wallets like Paytm, PhonePe, and Google Pay, which makes it easy for customers to pay for their orders using their preferred payment method. This feature provides a seamless payment experience for customers Ghosh and Saha (2017).

5.4.16 Restaurant Ratings and Reviews

These platforms allow customers to rate and review restaurants and dishes, which helps other customers make informed decisions while ordering. This feature also provides feedback to restaurants, which helps them improve the quality of their food and service.

5.4.17 Advertisements

Zomato is known for its witty and humorous advertising campaigns. The brand's ad campaigns revolve around the theme of foodies and how food brings people together. One of their most popular campaigns was the "Ordering food without Zomato? It's like trying to fly a plane without a cockpit" campaign. Zomato also uses social media influencers and bloggers to promote its brand, and it sponsors food events and festivals to get its name out there.

when it comes to Swiggy focuses on creating emotional connections with their audience through their advertising campaigns. They use relatable scenarios to portray their brand as the solution to everyday problems. For example, their "What a delivery" campaign showed people who had a long day at work finally getting to enjoy a delicious meal delivered by Swiggy. Swiggy also partners with popular restaurants to offer exclusive deals and promotions to their customers.

Foodpanda is known for its quirky and innovative advertising campaigns. They often use animations and cartoons in their ads, making them memorable and easy to remember. One of their most popular campaigns was the "Don't cook, just Foodpanda" campaign, which encouraged people to order in instead of cooking at home. Foodpanda also offers cashback and discounts to incentivize their customers to keep ordering from their platform.

Faasos is a popular fast-food chain that has transitioned into the online food delivery space. They focus on promoting their products as convenient and delicious. Faasos uses bright colors and mouth-watering imagery in their ads, which appeals to their target audience. They also offer discounts and loyalty programs to retain customers.

Overall, these platforms are constantly introducing new features and innovations to enhance the customer experience and stay ahead of the competition.

5.5 Methodology

The aim of the study is to analyze the marketing strategies in food delivery services such as Zomato, Swiggy, FoodPanda, and Faasos. In this study, we compare the strategies of selected food delivery services based on the following parameters:

- Capitalization and size
- Market growth and competition
- Tie-ups with restaurants
- Customer loyalty
- Fleet size and management
- Employees

5.5.1 Data Collection

Data can be collected through secondary data that can be obtained from published sources such as industry reports, Blogs, academic journals, Articles, FY reports from the companies, and some government reports. The research is totally based on secondary data collection.

5.5.2 *Data Analysis*

In this analysis of data, the business growth strategies are analyzed using different methods of analysis:

- SWOT analysis
- Analysis of business models
- Analysis of marketing strategies

5.6 Results

In recent years, food delivery has become a highly competitive industry. Various players have entered the market and are offering their services to customers. However, the market is dominated by four major players—Zomato, Swiggy, Foodpanda, and Faasos. These companies have their unique way of doing business. In this article, we will compare these companies based on different parameters.

Capital infusion and size: Zomato and Swiggy are the two biggest players in the market with a significant market share. They have raised millions of dollars in funding, and this has helped them expand their operations across the country. On the other hand, Foodpanda has not been able to raise as much capital as its competitors. It was recently acquired by Ola, and its operations have been merged with Ola's food delivery business. Faasos is a smaller player, but it has been able to attract funding from venture capitalists.

Market growth and competition: Zomato and Swiggy have been expanding aggressively and have been able to capture a significant share of the market. Foodpanda, on the other hand, has been struggling to keep up with its competitors. Faasos is a smaller player and is focusing on niche markets. The competition in the food delivery industry is intense, and companies are constantly innovating to stay ahead of their competitors.

Tie-ups with restaurants: Zomato and Swiggy have tie-ups with thousands of restaurants across the country. This allows them to offer a wide variety of cuisines to their customers. Foodpanda also has tie-ups with several restaurants, but its operations have been integrated with Ola's food delivery business. Faasos operates its kitchens and offers its signature dishes to its customers.

Customer loyalty: Zomato and Swiggy have a large customer base and have been able to build customer loyalty through their innovative products and services. Foodpanda has been struggling to retain its customers, and its acquisition by Ola has not helped its cause. Faasos is a smaller player, but it has been able to attract a loyal customer base through its unique offerings.

Fleet size and management: Zomato and Swiggy have large fleets of delivery personnel and have been able to manage them effectively. FoodPanda's fleet size is smaller, and its operations have been merged with Ola's food delivery business. Faasos operates its kitchens and has a smaller fleet of delivery personnel.

Employees: Zomato and Swiggy have a large workforce and have been able to attract and retain talent. FoodPanda's operations have been merged with Ola's food delivery business, and its employees have been integrated into Ola's workforce. Faasos has a smaller workforce, but it has been able to attract talent from the food and beverage industry.

Zomato and Swiggy are the two biggest players in the market, and they have been able to capture a significant share of the market. Foodpanda has been struggling to keep up with its competitors, and its acquisition by Ola has not helped its cause. Faasos is a smaller player, but it has been able to attract a loyal customer base through its unique offerings. The competition in the food delivery industry is intense, and companies are constantly innovating to stay ahead of their competitors.

Foodservice organizations increased their revenue growth by sticking to the marketing strategies outlined above.

5.6.1 Discussion

SWOT Analysis

	Strengths	Weakness	Opportunities	Threats
	Wide international presence, operating in over 24 countries	Relatively high delivery fees and commissions, leading to increased costs for customers and restaurant partners	Further expansion into new markets, both domestically and internationally, to increase revenue and user base	Increased regulatory scrutiny and legal challenges, as food delivery platforms come under increased scrutiny for labor practices and data privacy concerns
	Strong brand recognition and recall, with a well-designed user interface and marketing campaigns	Inconsistent quality of delivery service and customer service	Continued diversification of services and offerings beyond food delivery	Competition from other established players and new entrants into the market

(continued)

(continued)

	Strengths	Weakness	Opportunities	Threats
Zomato	Diversified offerings beyond food delivery, including restaurant reviews, reservations, and other food-related services	Heavy reliance on advertising and promotional campaigns to attract and retain customers	Expansion into adjacent markets, such as online grocery and other delivery services	Changing consumer preferences and habits, such as increasing health consciousness, leading to potential changes in demand for certain types of food
	Good relationships with restaurant partners, offering exclusive promotions and deals	Relatively high delivery fees and commissions, leading to increased costs for customers and restaurant partners		Increased regulatory scrutiny and legal challenges, as food delivery platforms come under increased scrutiny for labor practices and data privacy concerns
	Strong presence in the Indian market, with a wide range of food options available	Limited international presence, with a majority of their operations being limited to India	Expansion into new markets, both domestically and internationally, to increase revenue and user base	Increasing regulatory scrutiny and legal challenges, as food delivery platforms come under increased scrutiny for labor practices and data privacy concerns
	Highly user-friendly and efficient platform, with a fast and reliable delivery service	Heavy reliance on third-party logistics and delivery personnel, which can sometimes lead to quality control issues and delays	Partnering with more restaurants and offering exclusive deals and promotions to attract more customers	Competition from other established players and new entrants into the market
Swiggy	Strong brand recognition and recall, with popular advertising campaigns and promotional offers	Increased competition from other food delivery platforms, leading to a crowded marketplace and potential for loss of market share	Diversifying offerings beyond food delivery, such as grocery and medicine delivery, to expand the platform's reach and relevance	Changing consumer preferences and habits, such as increasing health consciousness, leading to potential changes in demand for certain types of food

(continued)

(continued)

	Strengths	Weakness	Opportunities	Threats
	Good relationships with restaurant partners, allowing for timely and consistent delivery	Limited international presence, with a majority of their operations being limited to India		Increasing regulatory scrutiny and legal challenges, as food delivery platforms come under increased scrutiny for labor practices and data privacy concerns
	Strong brand recognition and recall, with a wide range of restaurant partners and food options	Limited international presence, with operations mostly limited to Asia	Expansion into new markets, both domestically and internationally, to increase revenue and user base	Increased regulatory scrutiny and legal challenges, as food delivery platforms come under increased scrutiny for labor practices and data privacy concerns
Food panda	Good relationships with restaurant partners, offering exclusive promotions and deals	High delivery fees and commissions, leading to increased costs for customers and restaurant partners	Partnering with more restaurants and offering exclusive deals and promotions to attract more customers	Competition from other established players and new entrants into the market
	Diversified offerings beyond food delivery, including grocery and convenience store delivery	Inconsistent quality of delivery service and customer service	Diversifying offerings beyond food delivery, such as grocery and medicine delivery, to expand the platform's reach and relevance	Changing consumer preferences and habits
	User-friendly app and website: The app and website are easy to use, making it convenient for customers to order food			
	Strong brand recognition: Faasos has established a strong brand in the online food delivery market in India	Limited geographical presence: Faasos operates in a limited number of cities in India, which limits its customer base	Expansion into new cities: Faasos has the opportunity to expand its operations to new cities in India and increase its customer base	Intense competition: The online food delivery market in India is highly competitive, and Faasos faces stiff competition from established players like Swiggy and Zomato

(continued)

(continued)

	Strengths	Weakness	Opportunities	Threats
Faasos	Wide range of offerings: Faasos offers a wide range of cuisine options and has a diverse menu to cater to different taste preferences	High delivery charges: Faasos charges relatively higher delivery fees compared to its competitors, which can deter some price-sensitive customers	Diversification of offerings: Faasos can explore the possibility of diversifying its offerings to include grocery delivery, meal kits, and other food-related products and services	Changing Consumer preferences and behavior can change rapidly, and Faasos needs to constantly adapt to meet their changing needs
	Efficient delivery network: Faasos has a robust delivery network that ensures timely delivery of orders	Dependence on third-party delivery partners: Faasos relies on third-party delivery partners for last-mile delivery, which can impact the quality of delivery service	Partnerships with other brands: Faasos can collaborate with other brands to offer joint promotions and discounts to customers	Regulatory challenges: The Indian government has imposed regulations on the online food delivery industry, and Faasos needs to ensure compliance with these regulations to avoid penalties and legal issues
	Digital-first approach: Faasos has leveraged technology to offer a seamless online ordering experience through its website and mobile app			

Source Author's compiled data

5.6.2 Analysis of Business Model

Zomato: Zomato is an online food ordering and delivery platform that connects customers with restaurants. The business model of Zomato is based on a commission-based revenue model, where they earn a percentage of the order value from restaurants. Zomato also earns revenue through advertising and promotional activities, where restaurants can promote their menus and offerings to a wider audience. Additionally, Zomato offers a premium subscription service called Zomato Gold, which provides discounts and offers on selected restaurants and bars.

Swiggy: Swiggy is an on-demand food delivery platform that enables customers to order food from a wide range of restaurants. The business model of Swiggy is based on a commission-based revenue model, where they charge a commission from restaurants for every order placed through their platform. Swiggy also earns revenue through advertisements, where restaurants can promote their menus and offer to a larger audience. Additionally, Swiggy offers a premium subscription service called Swiggy Super, which provides free delivery, exclusive offers, and faster delivery to customers.

Foodpanda: Foodpanda is a food delivery platform that allows customers to order food from a variety of restaurants. The business model of Foodpanda is based on a commission-based revenue model, where they charge a commission from restaurants for every order placed through their platform. Additionally, Foodpanda offers a premium subscription service called Foodpanda Plus, which provides customers with exclusive discounts and offers on selected restaurants.

Faasos: Faasos is a food delivery platform that offers a range of food options to customers. The business model of Faasos is based on a cloud kitchen model, where they have their own centralized kitchens to prepare food for delivery. Faasos earns revenue through the sale of food items and delivery charges. Additionally, Faasos offers a subscription service called Faasos Elite, which provides customers with free delivery, exclusive offers, and faster delivery.

Analysis of marketing strategies: Analysis of the market strategies of Zomato, Swiggy, Foodpanda, and Faasos from the beginning of the company in food service marketing.

Zomato: Zomato has built a strong brand image by investing in a unique and memorable logo, consistent brand messaging, and a user-friendly website and mobile app. Zomato has focused on building a strong network of restaurant partners to offer a wide range of cuisine options to its customers. It also offers restaurant partners a range of marketing services to help them improve their visibility and attract new customers. Zomato's emphasis on user reviews and ratings has helped build a sense of community and trust. Overall, Zomato's marketing strategies have helped it become a leading player in the food service marketing industry in India.

Swiggy: Swiggy has focused on offering hyperlocal delivery services, which allowed them to offer fast and efficient food delivery within a limited radius. This helped them stand out from other food delivery services that were more focused on broader geographic areas. Swiggy leveraged technology to provide a seamless and efficient user experience. It used data analytics and machine learning to personalize the user experience, improve delivery times, and offer targeted marketing messages. Swiggy's partnerships with popular restaurants and exclusive deals and discounts for users have helped it build credibility and attract new customers.

Foodpanda: Foodpanda has positioned itself as a convenient food delivery option for busy customers. It offers a simple and easy-to-use mobile app and website, along with a range of payment options to make ordering food as simple as possible. Foodpanda has partnered with a wide range of restaurants, which has allowed them to offer customers a broad range of cuisine options. It also offers restaurant partners marketing services to help them improve their visibility and attract new customers. Foodpanda has used promotions and discounts to attract new customers and encourage repeat business, including first-time user discounts, referral bonuses, and loyalty programs.

Faasos: Faasos has focused on offering unique and innovative cuisine options that are not widely available in the market. It also offers a range of customizable options, which allows customers to tailor their meals to their specific tastes and preferences. Faasos leveraged technology to offer a seamless and efficient user experience, using data analytics and machine learning to personalize the user experience and offer targeted marketing messages. Faasos has offered a loyalty program to encourage repeat business and build a loyal customer base. It also offers exclusive deals and discounts to loyalty program members.

Overall, these companies have used a combination of technology-driven marketing, partnerships with restaurants, promotions, and discounts, and a focus on unique cuisine options to attract and retain customers. They have leveraged social media and online marketing to reach a wider audience and build brand awareness. These marketing strategies have helped these companies become major players in the highly competitive foodservice marketing industry in India.

5.7 Conclusion

In this conclusion, we researched and compared delivery services and their marketing strategies. We discovered that Zomato, in comparison to competitors such as Swiggy, Foodpanda, and Faasos, has a better marketing strategy and a more stable selection of available restaurants. Zomato's marketing strategies are unique and interesting, which is why so many people choose to use the service. They maintain a large selection since they have eateries across the price spectrum, from casual to upscale. Zomato is now a one-stop shop for all the foodies out there. Nevertheless, Swiggy, Foodpanda, and Faasos have a narrow focus and don't actively seek out new clients, limiting their potential market. As a result of its extensive marketing campaign and wide selection of options, Zomato has successfully established itself as the industry standard in the meal delivery space.

References

Shah A (2020) The effect of instagram influencers on vacation destination choice, pp 9–40

Bhotvawala MA, Bidichandani N, Balihallimath H, Khond MP (2016) Growth of food tech: a comparative study of aggregator food delivery services in india. In: Proceedings of the international conference on industrial engineering and operations management, pp 140–149

Choudhary N (2019) Strategic analysis of cloud kitchen–a case study. Manag Today 9(3):184–190. https://doi.org/10.11127/gmt.2019.09.05

A. I. Conference and M. Studies (2017) Academic international conference multidisciplinary studies and education, vol 2, no April 2017

Fatima NN (2022) Project of management information system (MIS207) Submitted By," no May, pp 0–14

Ghosh R, Saha TR (2018) A study of E-payment system on food delivery industry: a case study on swiggy. Int J Recent Trends Bus Tour 2(3):19–25. http://apps.ijrtbt.org/uploadedfiles/article_pdfs/7345b8bb-d777-488b-ae10-c1cf762ba3f5.pdf

Hole Y, Pawar S, Bhaskar MP (2018) Service marketing and quality strategies. Period Eng Nat Sci 6(1):182–196. https://doi.org/10.21533/pen.v6i1.291

Jha G, Singh P, Sharma L (2019) Recent advancements of augmented reality in real time applications. Int J Recent Technol Eng 8(2) Special Issue 7:537–542. https://doi.org/10.35940/ijrte.B10100.0782S719

Katsikeas C, Leonidou L, Zeriti A (2020) Revisiting international marketing strategy in a digital era: opportunities, challenges, and research directions. Int Mark Rev 37(3):405–424. https://doi.org/10.1108/IMR-02-2019-0080

Lau T, Ng D (2019) Online food delivery services: making food delivery the new normal. J Mark Adv Pract 1(1):17

Li C, Mirosa M, Bremer P (2020) Review of online food delivery platforms and their impacts on sustainability. Sustain 12(14):1–17. https://doi.org/10.3390/su12145528

Meenakshi N, Sinha A (2019) Food delivery apps in India: wherein lies the success strategy? Strateg Dir 35(7):12–15. https://doi.org/10.1108/SD-10-2018-0197

Mishra AK, Rath BN, Dash AK (2020) Does the Indian financial market nosedive because of the COVID-19 Outbreak, in comparison to after demonetisation and the GST? Emerg Mark Financ Trade 56(10):2162–2180. https://doi.org/10.1080/1540496X.2020.1785425

Raina A, Rana VS (2019) Popularity of online food ordering and delivery services-a comparative study between zomato, swiggy and uber eats in Ludhiana. Int J Manag Technol Eng 9(3):6080–6088. https://www.researchgate.net/publication/341109906

Singh L, Agrawal AK (2022) A case study on marketing strategy of boat. CASE J September 2011, 1–20. https://doi.org/10.1108/TCJ-06-2021-0087

Visnjic I, Neely A, Jovanovic M (2018) The path to outcome delivery: interplay of service market strategy and open business models. Technovation 72–73(December 2017):46–59. https://doi.org/10.1016/j.technovation.2018.02.003

Chapter 6
A Study on Consumer Behaviour Purchasing Pattern on Durable Products in Sustainable Consumer Market

G. Pradish Kumar and I. Shanthi Imaculate Jaculin

Abstract The large scale production of goods and the durability of usage by the consumer to enrich their life style. The innovation, technological advancement and changing consumer taste and preference led to more production of durable goods/ products. The consumers are led by more socially, economically and political in chose of purchase of product. The purchasing pattern of consumer are induced by earning capacity, socio-economic, cultural, demographic, etc. the emerging consumer are more vibrant and knowledgeable in choose of purchase. The migration of consumer from rural to urban area has led more demand for durable product. The changing life style pattern along with their migration shown enormous consumption of durable product. The growth rate of consumer durable on an average of 15%. The growing population is tender towards more white goods, modern kitchen appliance, electronics wearables. This present scenario paved to strong competition and large attraction towards Indian market. Now Indian market is become potential for global competitor to market their durable product. They are tapped by the new innovation and adaptability to new technology by the consumer.

Keywords Consumer behaviour · Durable · Purchase · Consumption · Market · Buying pattern

6.1 Introduction

The study of Consumer behaviour has gained it significant long before and the major research are made after globalization of economy in past two decades. This has widened the market and number of competitors at large. The earlier studies researcher and scholar has made interdisciplinary approach such as Social, Anthropology, Psychology and Economics. According to Kotler and Keller (2007, p. 151)

G. Pradish Kumar · I. Shanthi Imaculate Jaculin (✉)
St. Ann's College of Arts and Science, Tindivanam, Tamil Nadu, India
e-mail: santhiimmaculate@gmail.com

A. Edward J. et al. (eds.), *Digital Transformation for Business Sustainability*, Contributions to Environmental Sciences & Innovative Business Technology,
https://doi.org/10.1007/978-981-99-7058-2_6

"Consumer Behaviour is the study of how individuals, groups, and organizations select, use and dispose of goods, services, ideas or experiences to satisfy their needs and wants".

The consumer needs, preference, beliefs and attitude enables the marketer to understand the consumer behaviour buying pattern. The key issues of behavioural changes in buyer, they are.

- Consumer perception of product they feel and select among the alternatives.
- The factors influenced by the consumer (family, friends, relative, mass media, internet).
- What is the level of needs and wants on their buying decision.
- How they are motivated to buy a durable products.

The great shift in focus of potential consumer both in urban as well as rural consumer in India shows a significant change in trend recent years. The growth of internet and mass assess to durable products. The growth of single young population and rapid growth of nuclear family in urban and semi-urban has increased the sale of durable product. The marketer encourages the buyer on purchase of durable product on credit. The market segmented the consumer on basis of young working groups, and modern household buyer is increased to large extent.

The structure of this paper as follows: The following section explains the consumer behaviour in general and conceptual and it highlights. The second section explains the objective of the study which helps in developing in hypotheses and determinants. This is followed by methodology, research design that include data collection, respondents and the survey method is conducted for the study. After analysing the data and result is presented in the fourth section. The fifth section gives the recommendation on consumer purchase behaviour. The last section contains the conclusion, limitations and for recommendation for future research.

6.2 Review of Literature

The review of literature gives large insight on each and every steps of research process. The review gives previous study of work done by researcher and enables to induce for further study. The present study is done on consumer behaviour buying pattern towards consumer durables goods.

Castillo (2017) in his study considered key element for understanding the consumer behaviour is in the study "The Connection between Big Five Personality Traits, Customer Empowerment, and Customer Satisfaction in the Retail Sector", the emphasis was on determining how the Big Five Personality Dimensions of Personality and Customer Empowerment relate to each other.

James et al. (2016) in his study, sample data from Austria, Brunei, France, Hong Kong, the United Kingdom and the United States are used to assess consumer responses to situations of unethical consumer behaviour. It has been discovered that nationality has a key role in predicting how customers will react to certain dubious behaviours. Age and religion affiliation are proven to be key forecasters of consumer virtuous attitudes, although gender is not.

Naidu (2015) in his study revealed "An assessment of consumer awareness in rural marketplaces". The research assessed the level of awareness in India's rural markets. It highlights the availability of gold in this rapidly expanding market that has been developing continuously since the 1980s. FMCG products and durable goods now account for a sizable portion of the market, each contributing 53% to the overall market.

Erasmus and Meriam (2012) has reviewed the paradox of progress: choosing a large household appliance by an inexperienced consumer. The findings confirmed the hypothesis that, in the absence of adequate product knowledge, consumers with limited consumer socialization may rely on representative indicators of quality, such as price, brand name and store image. Unfortunately, the use of these substitute indicators of quality does not always imply informed, responsible purchasing behaviour.

6.3 Objective of the Study

- To comprehend the psychology of customer behaviour towards durability.
- To identify the purchasing behaviour of consumer of durables in Pondicherry region.
- To identify the post-purchase behaviour and satisfaction of consumer durable goods.

6.4 Research Methodology

Data collection method: This study has been carried out with the help of primary data and the data has been collected from interview schedule conducted from consumers. The secondary data was taken from various sources such as website and reports and complied as said by the study.

6.5 Sampling Design

The study used a convenience sampling technique. The researcher has conducted the area of study over Pondicherry region. 306 samples have been selected from 350, for the study after dropped out incomplete and inappropriate responses.

6.6 Hypotheses of the Study

Ho1: There is no substantial difference between the accused's demographic data and durables.

Ho1: There is no significant distinction between marketing mix and consumer characteristics of durable goods.

Ho1: There is no significant difference between timing of the purchase and consumer behaviour for durable goods.

Ho1: There is no significant distinction between post-purchase behaviour and band loyalty, image, advertisement, band value and consumer redresses and grievance of durable goods.

6.7 Analysis on Consumer Behaviour

This chapter analyses the consumer purchase behaviour in Pondicherry region. The data is collected to know the demographic variables of the general profile.

6.8 Demographic Particulars of the Informant

Table 6.1 shows the following analysis among 306 respondents, age reveals that 66.3% fall in the age group of less 30 years and generation have the influence by shopping habits. The gender of male buyer is more induced in purchasing of consumer durable. Marital status of respondents reveals that unmarried consumer buys more than the consumer are in relationship. Education of the respondents shows that graduate are trends to buys consumer durables more. Monthly of respondents reveals that middle class groups have more shopping habits of purchasing consumers than other class of groups. Family types shows that in joint family have the habit of buying more consumer durables. Residential area of respondents shows that most of consumers who lives in urban area have shops more than peoples live in rural area.

6.8.1 Consumers' Characteristic on Marketing Mix

The classification made on four P's of marketing mix and the observation. The marketing involves keys elements like product, price, place and promotion of goods. Descriptive statistics and ANOVA between the marketing mix and chosen demographic variables are used to analyse the data.

Table 6.2 shows that the marketing mix with monthly income of respondents with product, the variables like high quality, design and value there is significant difference

Table 6.1 Demographics information of the Respondents

Particulars	Classification	Number of respondents	Percentage
Age	Below 30 years	203	66.3
	31–60 years	93	30.4
	Above 60 years	10	3.3
Gender	Male	205	67.0
	Female	91	29.7
	Transgender	10	3.3
Marital status	Married	108	35.3
	Unmarried	193	63.1
	Widower	5	1.6
Educational qualification	ITI/Diploma	16	5.2
	UG/PG	223	72.9
	Professional	67	21.9
Occupation	Students	147	48.0
	Business	97	31.7
	Homemaker	57	18.6
	Retired	5	1.6
Monthly income	Below Rs.25,000	268	87.6
	Rs.25001–Rs.50,000	28	9.2
	Above Rs.50,000	10	3.3
Family type	Nuclear	222	72.5
	Joint	84	27.5
No. of dependents	Less than 2 members	54	17.6
	2-4members	199	65.0
	Above 4 members	53	17.3
Residential area of the respondents	Urban	164	53.6
	Semi-urban	56	18.3
	Rural	86	28.1

Source Primary data

in mean factor affecting consumer behaviour while purchasing the durable. Whereas there is no significant difference in mean factor affecting consumer behaviour not considers the convenience of durable product.

Table 6.3 shows that the marketing mix with family type of respondents, the variables like high quality, design and value there is significant difference in mean factor affecting consumer behaviour while purchasing the durable. Whereas there is no significant difference in mean factor affecting consumer behaviour not considers the convenience of durable product.

Table 6.2 Means, standard deviation and F value for Monthly income of the respondents

Marketing mix	Variables	F	Sig
Product	High quality	4.137	0.017
	Design	11.970	0.000
	Value	4.439	0.013
	Convenience	0.062	0.939
Price	Discounted price	4.048	0.011
	Comparison of price with alternative	26.231	0.000
	Sold at low price	0.051	0.832
Place	Wholesale	1.975	0.141
	Multi-channel	0.863	0.423
	Direct sale	5.610	0.004
	Retail	8.743	0.000
	Internet	9.528	0.000
Promotion	Direct Marketing	7.647	0.001
	Free gift	0.345	0.708
	Special offers	9.850	0.000
	Advertisement	0.618	0.540

Source Primary data

Table 6.3 Means, standard deviation and F value for family type of the respondents

Marketing mix	Variables	F	Sig
Product	High quality	3.914	0.049
	Design	5.497	0.020
	Value	12.717	0.000
	Convenience	0.070	0.791
Price	Discounted price	3.914	0.049
	Comparison of price with alternative	0.474	0.492
	Sold at low price	0.070	0.791
Place	Wholesale	0.333	0.564
	Multi-channel	2.451	0.118
	Direct sale	5.293	0.022
	Retail	0.569	0.451
	Internet	6.138	0.014
Promotion	Direct marketing	10.045	0.002
	Free gift	5.465	0.020
	Special offers	9.191	0.003
	Advertisement	0.067	0.795

Source Primary data

Table 6.4 Means, standard deviation and F value for residential area of the respondents

Marketing mix	Variables	F	Sig
Product	High quality	4.620	0.011
	Design	10.286	0.000
	Value	1.095	0.336
	Convenience	0.556	0.574
Price	Discounted price	4.620	0.011
	Comparison of price with alternative	2.749	0.066
	Sold at low price	0.556	0.574
Place	Wholesale	6.876	0.002
	Multi-channel	1.006	0.000
	Direct sale	6.479	0.002
	Retail	7.968	0.000
	Internet	0.920	0.400
Promotion	Direct marketing	14.501	0.000
	Free gift	3.392	0.035
	Special offers	5.852	0.003
	Advertisement	1.285	0.278

Source Primary data

Table 6.4 shows that the marketing mix with residential area of respondents, the variables like high quality, design and value there is significant difference in mean factor affecting consumer behaviour while purchasing the durable. Whereas there is no significant difference in mean factor affecting consumer behaviour not considers the convenience of durable product.

6.8.2 Post-purchase Behaviour of Consumer

Ten characteristics were used to assess customer behaviour towards durable products after the sale. Factor analysis using principal component approach and vari-max rotation were used to condense the variables into factors based on the responses provided by the randomly selected customers (Table 6.5).

By analysing the correlation between the variables, ten variables are broken down into smaller components. (Perception on Post-purchase behaviour consumer). In this circumstance, five components may explain the majority of the original data after 10 variables are condensed (Table 6.6).

Table 6.5 Initial eigenvalues of post-purchase behaviour

Factors	Initial eigenvalues		
	Eigenvalue	Percentage of variance	Cumulative percentage of variance
1	3.158	31.582	31.582
2	1.626	16.256	47.838
3	1.093	10.932	58.770
4	1.011	10.114	68.885
5	1.001	10.012	78.897

Source Computed data

Table 6.6 Factor scores of post-purchase behaviour of consumer

Factor	Components	Factor scores
Factor 1: Consumer redresses & grievance	Service during warranty period is absolutely free	0. 906
	After sale service after warrantee period is satisfactory	0.779
Factor 2: Brand image	I have attachment for the Brand that I use	0.0775
	I believe that this brand has no alternative	0.745
Factor 3: Brand advertisement	Advertisement give Product information	0.617
	Advertisement result in price hike	0.632
Factor 4: Brand value	I want others to realize the name of the brand	0.892
	I have positive feeling about the selected brand	0.911
Factor 5: Brand loyalty	I love to buy the same brand again, if I need it	0.438
	I will switch over brand if it fail to satisfy	0.639

Source Computed data

Significant differences between consumer's demographics with the post-purchase behaviour (PPB)

One way ANOVA test is used to test for significant differences in post-purchase behaviour of consumer durable goods based on factors such as age, gender, marital status, educational attainment, occupation, family type, residential location, and yearly income of the customer. The following null hypotheses are framed in order to be identified:

Null hypothesis H_0: There is no substantially different of age with post-purchase behaviour of consumer durable goods

PPB with Factor 1: H_0 that "there is no significant impact on consumer's age on Consumer Redresses & Grievance" is rejected. ('F' value is 44.854 and 19.179 and Value is 0.000 ($p < 0.05$)).

PPB with Factor 2: H_0 that "there is no significant impact on consumer's age on Brand Image" is rejected. ('F' value is 4.580 and 21.833 and Value is 0.011 ($p < 0.05$)).

PPB with Factor 3: H_0 that "there is no significant impact on consumer's age on Brand Advertisement" is rejected. ('F' value is 20.229 and 20.827 and Value is 0.000 ($p < 0.05$)).

PPB with Factor 4: H_0 that "there is no significant influence on consumer's age on Brand Value" is rejected. ('F' value is 8.959 and 26.339 and Value is 0.000 ($p < 0.05$)).

PPB with Factor 5: H_0 that "there is no significant influence on consumer's age on Brand Loyalty" is rejected. ('F' value is 20.737 and 13.580 and Value is 0.000 ($p < 0.05$)).

Null hypothesis H_0: There is no discernible gender difference in the post-purchase behaviour of durable consumer items

PPB with Factor 1: H_0 that "there is no significant impact on consumer's gender on Consumer Redresses & Grievance" is rejected. ('F' value is 25.839 and 19.179 and Value is 0.059 ($p < 0.05$)).

PPB with Factor 2: H_0 that "there is no significant impact on consumer's gender on Brand Image" is rejected. ('F' value is 8.826 and 10.658 and Value is 0.000 ($p < 0.05$)).

PPB with Factor 3: H_0 that "there is no significant impact on consumer's gender on Brand Advertisement" is rejected. ('F' value is 13.395 and 14.824 and Value is 0.000 ($p < 0.05$)).

PPB with Factor 4: H_0 that "there is no significant impact on consumer's gender on Brand Value" is accepted. ('F' value is 0.952 and 0.127 and Value is 0.387 ($p > 0.05$)).

PPB with Factor 5: H_0 that "there is no significant impact on consumer's gender on Brand Loyalty" is rejected.

Null hypothesis H_0: There is no significant change of educational qualification with post-purchase behaviour of consumer durable goods

PPB with Factor 1: the formulated hypothesis H_0 that "there is no significant influence on consumer's educational qualification on Consumer Redresses & Grievance" is rejected.

PPB with Factor 2: H_0 that "there is no significant influence on consumer's educational qualification on Brand Image" is rejected.

PPB with Factor 3: H_0 that "there is significant influence on consumer's educational qualification on Brand Advertisement" is accepted.

PPB with Factor 4: H_0 that "there is no significant influence on consumer's educational qualification on Brand Value" is rejected.

PPB with Factor 5: H_0 that "there is no significant influence on consumer's educational qualification on Brand Loyalty" is rejected.

6.9 Analysis and Interpretation

The study made by the research and analysis are given under the following heads.

- All patterns of income group of consumers are to be considered while manufacturing a product or goods.
- The awareness and proper education about the product should be given to the consumers.
- The sale promotion will be boosted on various credit facility available to consumer which enable increase in volume of sales.
- Brand image and Brand value gives more important in increase of the sale volume as well as retain the old and potential consumer. The manufactures have to maintain its standards.
- Post-purchase behaviour of consumer another important factor which the consumer are satisfied after purchase and services of product should be done and give better grievance on product.
- The younger generations are prone to shift brand and other age groups seek social status hence the manufacture has to maintain its brand reliability.

6.10 Conclusion

The study shows that the consumers durable are wider market. The product or goods should be available to all and world market seeks its potential outcome in Indian market. Our country has evident growth in internet and communication in recent past year. The marketers have started using the digital device which resulted in numerous marketing opportunities and marketing domain for advertisement both domestic and international. The consumers are directly connected to seller in real-time intermediary service by digital marketing app and communication tools. Social media accounted for 25% of all marketing spending in Q2 of 2020, an increase of 13% from 2019. Hence there is a substantial growth in purchase of durables goods.

The study is limited to certain measurement scale only. The research is covered to consumer behaviour of durable goods of white goods and brown goods only. There is a possibility of factor response bias and uncontrolled. Thus the study can be conducted on comparing the durables and Non-durables products. The brand is important factor where it was found in the study, so brand attitude of consumer behaviour can be focused on the future study.

References

Castillo (2017) The consumer behaviour is "The relationship between big five personality traits"

Deiss R, Henneberry R (2017) Digital marketing for dummies

Erasmus AC, Meriam M (2012) The paradox of progress: inexperience consumer's choice of major household appliance

Hoyer DW, Macinnis JD, Pieters R Consumer behaviour, 6th edn

Kotler P, Keller KL, Koshy A, Jha M (2015) Marketing management-a south Asian prespective. Pearson. https://www.adroll.com/blog/digital-marketing-in-2021-the-trends-to-look-out-for

Kotler, Kevin KL (2007) Marketing Management-Prentice Hall,14th edn

Mcneal JU, Yeh C (2016) Consumer reaction to unethical consumer behaviour scenarios are investigated using sample data from Austria, Brunei, France, Hong Kong, UK and USA

Naidu YKM (2015) An evaluation of consumer awareness in rural markets

Oliver RL (1997) Satisfaction: a behavioral perspective on the consumer. McGraw Hill, New York

Philip B (2019) Unified payment interface–impact of UPI in customer satisfaction. Res Guru Online J Multidiscip Subj 12(4):124–129

Selvi VD (2006) The changing scenario of the consumption, pattern of employed people. Indian J Mark 22–36

Sarkar A (2008) Latest trends in consumer buying behaviour in life style centers worldwide. Icfai J Manag Res VII(6):70–82

Chapter 7
A Study of Financial Literacy as a Sustainable Factor Among Micro-entrepreneurs in Coimbatore District of Tamil Nadu

P. M. Varshini, Aravind Sakthivel, and K. N. Sakthivel

Abstract Financial literacy is perceived as a more pressing issue in wealthy countries than in developing ones, particularly after the occurrence of financial crises. It has become increasingly crucial to achieve financial stability and prevent financial catastrophes. Therefore, the focus of this research is to determine the level of awareness among microbusiness owners in the Coimbatore area of Tamil Nadu regarding the importance of becoming financially literate. To accomplish the study's objectives, a descriptive cross-sectional design was employed, utilising a survey approach with a structured questionnaire to assess financial literacy. The questionnaire aimed to measure the degree of awareness of financial education among micro-entrepreneurs. The data collected was subjected to analysis using both descriptive and inferential statistics. The survey findings revealed that microbusiness owners in the Coimbatore area are enthusiastic about contributing to financial literacy initiatives. This indicates that they are well aware of the value and potential advantages offered by being financially literate.

Keywords Financial literacy · Micro entrepreneurs · Small entrepreneurs

---

P. M. Varshini
Tapasya PU College, Bangalore, India

A. Sakthivel
Natwest Group, Bangalore, India

K. N. Sakthivel (✉)
Green Board Educators, Bangalore, India
e-mail: knsvel2000@gmail.com

A. Edward J. et al. (eds.), *Digital Transformation for Business Sustainability*,
Contributions to Environmental Sciences & Innovative Business Technology,
https://doi.org/10.1007/978-981-99-7058-2_7

7.1 Introduction

The effects of the world's financial crisis prompted policymakers all around the world to focus on financial literacy (Hasibuan et al. 2017). Financial literacy educates consumers and suppliers of financial goods about the risk and return of such products (Sibi 2021). To prevent financial crises by achieving financial stability, financial literacy has become a must. This understanding of financial issues aids in risk hedging and preserving the stability of the financial systems. Better understanding of the components of financial literacy and investment behaviour is thus crucial for effective development strategy (Shruthi et al. 2018). Financial literacy is defined by the Organization for Economic Cooperation and Development (OECD) as "the accumulation of financial knowledge, information, abilities, mentality, and behaviours required to achieve wise financial choices and, eventually, achieve sustainable fiscal viability" (Singhal et al. 2015). Financial literacy is a more pressing problem in wealthy countries than in developing ones. It is an essential instrument for addressing the problem of information asymmetry between financial organisations and those who use their services (Kalaiselvi and Krishnan 2021). It makes it easier for customers to understand and use things. Programmes for financial education are tailored to each client's specific financial needs. These help customers comprehend financial tools via which they may address their worries about managing risk and cash flow. It also gives consumers more power by improving their knowledge of financial goods and how to use them. Their level of literacy enables people to exercise their rights in the context of consumer protection (Devi 2020).

Programmes for financial literacy are developed to be applicable for each societal group. It assists customers in using a range of financial services throughout the official and informal financial sectors (Timcy 2018). Financial goods and services are in demand because financial literacy raises public knowledge of them. It quickens the speed of financial inclusion by helping individuals comprehend the demands for and advantages of the items and services provided by banks (Suganthi and Asokhan 2022a).

The Reserve Bank of India launched the financial literacy programme using Karnataka as a pilot state in order to fulfil its goals and meet the need for financial literacy. The Reserve Bank of India launched this financial literacy programme to improve the lives of the general public and to help them comprehend the institution's duties (Development 2018). The programme included events including displaying information about financial goods and services, allowing regular people to connect with bankers, and encouraging the use of information and communication technology (ICT). Presentations, brochures, leaflets, videos and other forms of information dissemination were used to reach the target audience (Siregar 2022). The RBI updated the rules for Financial Literacy Centers (FLC) in 2012. The instructions recommended commercial banks to step up their efforts to promote financial literacy via FLC and all of its rural bank locations. A minimum of once a month should be set aside for financial literacy camps (Sangwan et al. 2020). Financial inclusion should be promoted by the two fundamental principles of "Financial Literacy" and

"Financial Access." In addition to everything else, the Reserve Bank has aggressively engaged commercial banks, non-governmental organisations, government apparatus, schools, universities and other institutions to carry out the programme (Suganthi and Asokhan 2022b).

The Securities and Exchange Board of India has also undertaken ongoing efforts to promote financial literacy and disseminate information about financial markets. SEBI has run a number of initiatives for the various societal sectors in an effort to increase the number of investors who take part in India's economic development (Andarsari and Ningtyas 2019). Along with the groundwork, SEBI also developed a website to serve the demands of investors in an effort to incorporate all facets of society. SEBI has acknowledged Sanchayan as an investor association for its work with financial literacy initiatives (Suthamathi 2018).

Additionally, the Financial Literacy Advisory Board India was established by the American Academy of Financial Management India (AAFM India) (F-LAB India). With the goal of establishing a financially aware and empowered India, the board was constituted in accordance with Section 25 of the Companies Act of 1956. Low levels of financial literacy have been noted in developing nations (Ayhan 2019). According to data from India, more than half of the surveyed workers keep their cash at home and take out loans at higher interest rates. India was placed 23rd out of 28 economies in the VISA 2012 Global Financial Literacy Barometer, according to a survey on global financial literacy (Rachel Kuruvilla and Harikumar 2020). The survey showed that while Indians have the greatest rates of saving among their international counterparts and are extremely cognizant of their savings, household investing knowledge is quite poor. The research paper is structured into five main sections to systematically explore the topic of financial literacy among microbusiness owners in the Coimbatore area of Tamil Nadu. The first section is the introduction, which provides an overview of the significance of financial literacy, especially in the aftermath of financial crises, and sets the context for the study with the objectives of the study. The second section comprises the literature review, where existing research and relevant theories related to financial literacy and its impact on microbusiness growth are thoroughly examined. The third section outlines the methodology employed in the research, presenting the descriptive cross-sectional design and the use of a structured questionnaire to assess financial literacy among the respondents. Moving on to the fourth section, the results and discussion are presented, divided into two subsections: 4.1 examines the financial literacy levels of the respondents, while 4.2 focuses on measuring the growth of micro and small enterprises (MSEs). Finally, the fifth section provides a comprehensive conclusion that summarises the findings and emphasises the importance of financial literacy for microbusiness owners, along with implications for future initiatives and policies to enhance financial literacy in the Coimbatore region. Overall, this well-structured study aims to shed light on the awareness and significance of financial literacy among microbusiness owners and its potential impact on the growth of MSEs in the area.

7.1.1 *Objectives of the Study*

The study's primary goal is to ascertain the association between financial literacy and the expansion of micro & small businesses in the Coimbatore region of Tamil Nadu.

The study addressed the following specific objectives:

- To ascertain how debt management literacy affects SME development in the Coimbatore region of Tamil Nadu.
- To determine how banking literacy influences the development of MSEs in the Coimbatore region of Tamil Nadu.
- To investigate how bookkeeping literacy affects the development of MSEs in the Coimbatore region of Tamil Nadu.

The goal of this study is to evaluate the degree of financial literacy among those running microbusinesses in the Coimbatore area. For the purpose of accomplishing the objectives of the investigation, a description cross-sectional study was carried out.

7.2 Literature Review

Financial literacy among Cypriot individuals was surveyed in Andreou and Anyfantaki (2021), and reported on the findings as well as the financial behaviour of the respondents. Also, it investigates for the first time how respondents' financial literacy affects how they utilise online banking and other digital financial services. The results suggested that there is a relationship that is statistically substantial among the degrees of financial literacy and the quantity of usage of online banking. The analysis revealed how digital competence and financial literacy interact, with implications for how people utilise i-banking services.

Prasad et al. (2018) had analysed the amount of knowledge that households have about their finances in India's Rajasthan state's Udaipur city. Knowledge of a variety of digital platforms and how to make use of them is an essential component of financial literacy. The research also determined the influence of personality traits on financial literacy in the digital age. Data on digital financial literacy were gathered by surveying people and using a well-designed questionnaire. The findings of the research will offer a valuable pathway for both providers of digital platforms and the government in their efforts to encourage citizens to engage in digital transactions. The study also recommended that in order to expand the reach of digital transactions, there is a need for a wave of awareness campaigns.

Gupta and Kaur (2014) had analysed the degree of financial literacy knowledge among the microbusiness owners in Himachal Pradesh's district Kangra. To meet the goals of the research, a survey approach was utilised. A structured questionnaire was developed as a tool to assess financial literacy. Micro-entrepreneurs' knowledge of financial education was examined using descriptive and inferential analysis.

Rachel Kuruvilla and Harikumar (2020) determined the link between a woman's educational background and her degree of financial knowledge. The results show a relationship between an entrepreneur's educational background and the marks they received on the exam used to gauge their financial understanding.

Anshika and Mallik (2021) assessed the influences on the overall, specific and general levels of financial literacy among MSE founders. In addition, the findings show that service-providing micro- and small-sized enterprises often have a far higher degree of financial literacy than their manufacturing sector peers. The poll also found that business owners who have completed graduate school scored much better in every category of financial literacy. Gross profit ratio is the most reliable indicator of an entrepreneur's financial acumen, according to the research.

Tuffour et al. (2022) the goal of this research is to find out how much of an impact managers' financial literacy (awareness, attitude and knowledge) has on the growth of small businesses in Ghana's La Nkwantanang Madina Municipality. The findings suggested that workers' financial knowledge had a major impact on a business's bottom line (both financial and non-financial performance). In addition, the three facets of financial literacy—awareness, attitude and knowledge—make significant contributions to enhanced performance across a wide range of financial and non-financial sectors. It is recommended to increase financial literacy among small company managers and owners by implementing capacity building activities.

7.3 Methodology

The purpose of this research is to assess the level of financial knowledge among microbusiness owners in the Coimbatore area. This research used a descriptive cross-sectional design to accomplish its objectives (Thomas and Subhashree 2020).

Primary and secondary data were utilised to complete this investigation. The major data utilised in this research was gathered by the administration of a structured questionnaire that included both closed- and open-ended items. The closed-ended questions were designed using a Likert scale that consisted of five to six elements. Analysis of government documents, papers, journal articles and industry reports provided secondary data on the impact of financial education on the development of MSEs, as well as the status of financial education initiatives now underway in Coimbatore (Kulathunga et al. 2020). To determine the value of the assets, specific MSE business documents were also looked at while they were accessible. The research had a sample size of 200, and the target population consisted of company owners in MSEs that were enrolled with Revenue Department under an individual business permit and had less than 20 employees. The demographic data acquired revealed 114 males and 86 females, as well as their education level (Primary, secondary, diploma) and type of business (Manufacturing, Retail/wholesale, Service and Agribusiness).

We assessed the data and built the models using statistical tools and methodologies (Ye and Kulathunga 2019).

Using SPSS Version 26.0, after encoding, the information was input into the computer. The analysis and summarization of quantitative data was done using means, frequencies and percentages. On qualitative data, counting and content analysis were performed. The growth of MSEs and financial literacy were examined and a cross-tabulation and frequency analysis was carried out in order to investigate the degree of correlation that existed between the two sets of data (Mohd Thas Thaker et al. 2021).

7.4 Results and Discussion

The analysis and research results are presented in this chapter. The findings are discussed in relation to the growth of MSEs and financial literacy.

According to the results of the study found in the research, the majority of the participants in the study were males, (57.0%) of the respondents and females (43.0%) of the participants as may be seen in Table 7.1.

Most respondents (50.5%) said they were between the ages of 36 and 60, followed by 28.5% said they were between the ages of 18 to 35 and 21.0% said they were older than 60. This indicates that, as shown in Table 7.2, the vast majority of MSE managers are adults, and their ages vary from 36 to 60.

According to the information in Table 7.3, the vast majority of those who responded (45.0%) had a secondary education, followed by 19.5% with a primary education, 23.5% with a college degree (a diploma) and 12.0% without any formal education. This indicates that the majority of the 76.5% of MSE managers in the research region had less education than a college degree.

Table 7.1 Gender of respondents

	Frequency	Percent
Male	114	57.0
Female	86	43.0
Total	200	100.0

Source Primary data

Table 7.2 Age of respondents

	Frequency	Percent
18–35	57	28.5
36–60	101	50.5
Above 60	42	21.0
Total	200	100.0

Source Primary data

Table 7.3 Education level of the respondents

	Frequency	Percent
Primary	39	19.5
Secondary	90	45.0
Diploma	47	23.5
None	24	12.0
Total	200	100.0

Source Primary data

Table 7.4 Type of business of the respondents

	Frequency	Percent
Manufacturing	35	17.5
Retail/wholesale	91	45.5
Service	46	23.0
Agribusiness	28	14.0
Total	200	100.0

Source Primary data

In accordance with Table 7.4, a large proportion of the individuals who responded (45.5%) work in the retail or wholesale sector, followed by the service sector (23.1%), the agribusiness (14.0%) and the manufacturing sector (17.5%). Due to the minimal capital requirements for the industry, the majority of MSEs participate in wholesale and retail business.

According to Table 7.5, the majority of respondents (42.5%) for the 6–10 h of duration in business, followed by the duration of 0–5 h. (26.0%), the duration between 11–15 h (17.5%) and the duration between 16–20 h (14%). From this table, we can conclude that majority of the respondents agreed for working in between 6 and 10 h.

Table 7.5 Duration in business

	Frequency	Percent
0–5	52	26.0
6–10	85	42.5
11–15	35	17.5
16–20	28	14.0
Total	200	100.0

Source Primary data

7.4.1 Financial Literacy of the Respondents

The respondents' level of financial literacy was evaluated based on their knowledge of the following four primary subscales: Literacy in debt, budgeting, banking services and bookkeeping are all important aspects of financial literacy.

Six questions on a Likert scale were given to respondents to indicate their degree of agreement with the statement. According to Table 7.6, the modal answer for whether respondents paid loan payments on time was agree (65.5%), strongly agree (13.5%), disagree (12.0%), strongly disagree (7.0%) and neutral (2.0%). This indicates that the majority of MSE owners were successful in repaying their loans on time. The ability to compare terms and conditions before purchasing financial products was the theme of the following question. The modal answer for this was agree (56.5%), with 20.0% of respondents strongly agreeing, 7.0% disagreeing strongly, 15.5% disagreeing and 1.0% neutral. This indicates that the majority of consumers, before to making purchases, were examining the terms and conditions offered by a variety of lenders. From the above table, it is clear that 49.0% of the respondents either strongly disagreed (10.0%) or disagreed (39.0%) with the statement that firms utilize more than half of the income to repay loans, while a total of 48.5% of the respondents either agreed or strongly agreed with the statement. This demonstrates that the majority of firms used loans inside their operations that were within their budgets. 52.5% of the respondents disagreed with the statement Access to Various Sources of Finance by (13.5%) strongly disagree and (39.0%) disagree, the agree percentage for this statement is 29.0% and 18.5% are Neutral. The other statement "Effect of Inflation and Interest Rates" 28% of respondents strongly disagreed and 65% of them responded Disagreed. Whereas, 19% of respondents are neutral and 78% of respondents agreed and 10% of respondents are strongly agreed. The statement "Can Determine Accurately the Total Debt Position of my Business" 7.5% of respondent's strongly disagreed and 21.5% of respondents had disagreed. Whereas, 16% of respondents are neutral and 48.5% of respondents agreed and 6.5% of respondents are strongly agreed. This table indicates 78% of respondents are disagreed to "Use More than Half Business Revenue to Repay Loans" statement.

Table 7.7 presents budgeting in business is a process of looking at a business' estimated incomes. We can see from the results that 41% of respondents never prepare written annual financial objective and 20% of respondents rarely prepare annual financial objective. 27.5% of respondents sometimes prepare their financial objective, whereas 7% of respondents often prepare their financial objective. 4.5% of respondents always prepare written annual financial objective. For the second statement "prepare annual budgets for the business" 46.5% of respondents never prepared any annual budgets for the business and 14.0% of respondents rarely prepared annual budgets for the business. Whereas, 24.0% of respondents sometimes prepare annual budgets for the business as well as 11.5% of respondents often prepare annual budgets and 4.0% of respondents always prepare annual budgets for the business. Third statement "ability to compare financial objectives and business performance" 51.5% of respondents never compare financial objectives and business performance and

Table 7.6 Table of debt management literacy

		Strongly disagree	Disagree	Neutral	Agree	Strongly agree	Total
Timely loan payment	Frequency	14	24	4	131	27	200
	Percent	7.0	12.0	2.0	65.5	13.5	100.0
Ability to compare terms and condition	Frequency	14	31	2	113	40	200
	Percent	7.0	15.5	1.0	56.5	20.0	100.0
Use more than half business revenue to repay loans	Frequency	20	78	5	70	27	200
	Percent	10.0	39.0	2.5	35.0	13.5	100.0
Access to various sources of finance	Frequency	27	78	37	43	15	200
	Percent	13.5	39.0	18.5	21.5	7.5	100.0
Effect of inflation and interest rates	Frequency	28	65	19	78	10	200
	Percent	14.0	32.5	9.5	39.0	5.0	100.0
Can determine accurately the total debt position of my business	Frequency	15	43	32	97	13	200
	Percent	7.5	21.5	16.0	48.5	6.5	100.0

Source Primary data

14.0% of respondents rarely compare financial objectives and business performance. Whereas, 24.0% of respondents sometimes compare and 8.0% of respondents often compare, 4.0% of respondents always have the ability to compare financial objectives and business performance. Fourth statement "use budgets to set targets for employees" for this statement 51.5% of respondents never use budgets to set targets for employees and 15.5% of respondents rarely use budgets to set targets for employees. Whereas, 21.5% of respondents of respondents sometimes sets targets and 5.0% of respondents often sets targets. 6.5% of respondents always use budgets to set targets for employees. Six statement "Use Budgets to Track Spending" of their businesses for this statement 35.5% of respondents never use budgets to track spending and 20.5% of respondents rarely use budgets to track spending. Whereas, 25.0% of respondents sometimes use budgets to track spending and 11.5% of respondents often use budgets to track spending, 7.5% of respondents always use budgets to track spending.

Table 7.8 shows that 89.5% of respondents have said yes and 10.5% said no for operating a business bank account. 27.5% of respondents have said yes and 72.5% said no for banks business proceeds daily. 86.0% of respondents have said yes and 14.0% said no to get extra finance for the business form somewhere. 19.5 of respondents have said yes and 80.5% said no for awareness of commercial bank requirements of lending to MSEs. 49.0% of respondents have said yes and 51.0% said no for checking commercial bank websites to see their credit products and their

Table 7.7 Budgeting literacy among entrepreneurs

		Never	Rarely	Sometimes	Often	Always	Total
Prepare written annual financial objective	Frequency	82	40	55	14	9	200
	Percent	41.0	20.0	27.5	7.0	4.5	100.0
Prepare annual budgets for the business	Frequency	93	28	48	23	8	200
	Percent	46.5	14.0	24.0	11.5	4.0	100.0
Ability to compare financial objectives and business performance	Frequency	103	28	45	16	8	200
	Percent	51.5	14.0	22.5	8.0	4.0	100.0
Use budgets to set targets for employees	Frequency	103	31	43	10	13	200
	Percent	51.5	15.5	21.5	5.0	6.5	100.0
Use budgets to track spending	Frequency	71	41	50	23	15	200
	Percent	35.5	20.5	25.0	11.5	7.5	100.0

Source Primary data

Table 7.8 Banking services among entrepreneurs

		Yes	No	Total
Operates a business bank account	Frequency	179	21	200
	Percent	89.5	10.5	100.0
Banks business proceeds daily	Frequency	55	145	200
	Percent	27.5	72.5	100.0
Knows where to get extra finance for the business	Frequency	172	28	200
	Percent	86.0	14.0	100.0
Is aware of commercial bank requirements of lending to MSEs	Frequency	39	161	200
	Percent	19.5	80.5	100.0
Checks commercial bank websites to see their credit products and their requirements	Frequency	98	102	200
	Percent	49.0	51.0	100.0
Can access other bank products other than loans	Frequency	53	147	200
	Percent	26.5	73.5	100.0

Source Primary data

requirements. 26.5% of respondents have said yes and 73.5% said no to access other bank products other than loans. From the table, we can conclude that 86.0% of respondents know where to get extra finance for the business.

Table 7.9 presents the measure of Knowledge and Skills of the respondents about their business and financial matters related to the business. 34.5% of the respondents have no knowledge and skills to prepare financial statements and 24.0% of the

Table 7.9 Book keeping literacy

Statements		No knowledge and skills	Unsatisfactory knowledge and skills	Satisfactory knowledge and skills	Good knowledge and skills	Excellent knowledge and skills	Total
Ability to prepare financial statements	Frequency	69	48	50	29	4	200
	Percent	34.5	24.0	25.0	14.5	2.0	100.0
Ability to perform financial analysis	Frequency	73	55	44	28	0	200
	Percent	36.5	27.5	22.0	14.0	0	100.0
Maintenance of cashbook	Frequency	73	46	51	28	2	200
	Percent	36.5	23.0	25.5	14.0	1.0	100.0
Filling tax returns	Frequency	89	51	34	24	2	200
	Percent	44.5	25.5	17.0	12.0	1.0	100.0
Maintain and balance the ledger accurately	Frequency	85	52	34	26	3	200
	Percent	42.5	26.0	17.0	13.0	1.5	100.0
Ability to do bank reconciliations	Frequency	87	51	33	25	4	200
	Percent	43.5	25.5	16.5	12.5	2.0	100.0

Source Primary data

respondents have unsatisfactory knowledge and Skills to Prepare Financial Statements. Whereas, 25.0% of the respondents have satisfactory knowledge and skills, as well as 14.5% have good knowledge and skills. 2.0% have excellent knowledge and skills to prepare financial statements. 36.5% of the respondents have no knowledge and skills to Perform Financial Analysis and 27.5% of the respondents have unsatisfactory knowledge and Skills to Perform Financial Analysis. Whereas, 22.0% of the respondents have satisfactory knowledge and skills, as well as 14.0% have good knowledge and skills to Perform Financial Analysis. 36.5% of the respondents have no knowledge and skills to Maintenance of Cashbook and 23.0% of the respondents have unsatisfactory knowledge and Skills to Maintenance of Cashbook. Whereas, 25.5% of the respondents have satisfactory knowledge and skills, as well as 14.0% have good knowledge and skills. 1.0% of respondents have excellent knowledge and Skills to Maintenance of Cashbook. 44.5% of the respondents have no knowledge and skills Filling Tax Returns and 25.5% of the respondents have unsatisfactory knowledge and Skills to Filling Tax Returns. Whereas, 17.0% of the respondents have satisfactory knowledge and skills, as well as 12.0% have good knowledge and skills. 1.0% of respondents have excellent knowledge and Skills to Filling Tax Returns. 42.5% of the respondents have no knowledge and skills to Maintain and Balance the Ledger Accurately and 26% of the respondents have unsatisfactory knowledge and Skills to Maintain and Balance the Ledger Accurately. Whereas, 17.0% of the respondents have satisfactory knowledge and skills, as well as 13.0% have good knowledge and skills. 1.50% of respondents have excellent knowledge and Skills to Maintain and Balance the Ledger Accurately. 43.5% of the respondents have no knowledge and skills to Ability to do Bank Reconciliations and 25.5% of the respondents have unsatisfactory knowledge and Skills Ability to do Bank Reconciliations. Whereas, 16.5% of the respondents have satisfactory knowledge and skills, as well as 12.5% have good knowledge and skills. 2.0% of respondents have excellent knowledge and Skills to do Bank Reconciliations. From this table, we can conclude that 43.5% of the respondents have no knowledge and skills to Ability to do Bank Reconciliations.

7.4.2 Measuring MSE Growth

Based on the value of assets and the number of workers, MSE growth in the research region was measured.

Table 7.10 shows the value of assets; from the table it is concluded that 76.5% of the MSEs have the asset value of less than 1,00,000, only 5.5% have asset value of over 500,000/-.

According to Table 7.11's findings, MSEs' asset growth is positively impacted by their understanding of debt management. In particular, of the 42 MSEs that strongly disagreed, disagreed, or were unsure on timely loan repayments, only 13 have an asset value above 100,000/ =. In contrast, of the 21 MSEs that have grown their assets beyond 500,000/ =, only 2 disagreed that they are unable to repay their loan instalments on time. This indicates that timely loan repayments increase the owner

Table 7.10 Value of assets

		Frequency	Percent
Valid	500 < 50,000	87	43.5
	51,000 < 100,000	66	33.0
	101,000 < 500,000	36	18.0
	501,000 < 1,000,000	10	5.0
	Over 1000,000	1	0.5
	Total	200	100.0

Source Primary data

manager's access to development financing for business growth. Regarding the ability to compare terms and conditions, 47 MSE owners were strongly disagreed (14) or disagreed (31) or were undecided (2). As a result, only 6 out of 47 have assets valued at more than 500,000/-. This indicates that the inability to compare terms and conditions has a negative effect on asset growth since the owner manager may not choose the best goods for his business. On the question of whether MSEs utilize more than half of their business's earnings to pay back loans, 98 MSEs either strongly disagreed (20) or disagreed (78). Even though this suggests that the majority of firms used affordable loans, only 8 out of 98 have seen their assets increase by more than 500,000 in value. This may be due to MSEs' limited access to significant amounts of development finance as a result of low maximum loan limitations. For the Effect of Inflation and Interest Rates, 98 MSEs either strongly disagreed (13) or disagreed (19). As a result, 9 were neutral and 37 were agreed and (6) Strongly agreed. Regarding the Access to Various Sources of Finance (10) Strongly Disagreed for the asset valuation of under 100,000/-. (56) Disagreed for the asset valuation of under 50,000/-. 23 MSEs agreed and 7 strongly agreed for the asset valuation of under 50,000/-. The next statement regarding Determining Accurately the Total Debt Position of my Business 9 strongly disagreed and were under 100,000/-. 23 disagreed and 10 were neutral. 41 agreed and 13 strongly agreed for Determining Accurately the Total Debt Position of business under the asset valuation of less than 50,000/-.

The results that may be seen in Table 7.12 demonstrate that there is a causal link between the ability to create and stick to a budget and the increase in the value of assets held by MSEs. Specifically, the assets valued at less than 50,000/-, 60 MSEs never prepared annual budgets for the business for the company assets below 100,000/-. 12 MSEs rarely prepared annual budgets for the business. 20 MSEs sometimes prepared the annual budgets. The company assets of below 100,000/-. 6 MSEs of them often prepare the annual budgets. 70MSEs below 50,000/- never had the Capability to Evaluate Business Performance in Relation to Financial Goals. For the Use Budgets to Set Targets for Employees the company assets fewer than 50,000/-. 70MSEs never used Budgets to Set Targets for Employees. For the Make use of budgets to keep track of your spending. 53MSEs were never and 17 MSEs rarely under the company assets of 50,000/-. Whereas, 24 companies sometimes Use Budgets to Track Spending comes under the valuation of 100,000/-. For the 500,000/- company

Table 7.11 Debt literacy and growth in value of assets

		Value of assets					Total
		500 < 50,000	51,000 < 100,000	101,000 < 500,000	501,000 < 1,000,000	Over 1000,000	
Timely loan payment	Strongly disagree	5	3	2	4	0	14
	Disagree	10	8	4	2	0	24
	Neutral	2	1	0	0	1	4
	Agree	60	47	21	3	0	131
	Strongly agree	10	7	9	1	0	27
Total		87	66	36	10	1	200
Ability to compare terms and condition	Strongly disagree	4	2	4	4	0	14
	Disagree	16	9	5	1	0	31
	Neutral	0	1	0	0	1	2
	Agree	51	39	21	2	0	113
	Strongly agree	16	15	6	3	0	40
Total		87	66	36	10	1	200
Use more than half business revenue to repay loans	Strongly disagree	7	4	4	5	0	20
	Disagree	63	10	2	3	0	78
	Neutral	2	2	0	0	1	5
	Agree	5	40	23	2	0	70
	Strongly agree	10	10	7	0	0	27
Total		87	66	36	10	1	200
Effect of inflation and interest rates	Strongly disagree	6	9	13	0	0	28
	Disagree	31	19	15	0	0	65
	Neutral	7	9	3	0	0	19

(continued)

Table 7.11 (continued)

		Value of assets					Total
		500 < 50,000	51,000 < 100,000	101,000 < 500,000	501,000 < 1,000,000	Over 1000,000	
	Agree	37	26	5	9	1	78
	Strongly agree	6	3	0	1	0	10
Total		87	66	36	10	1	200
Access to various sources of finance	Strongly disagree	9	10	6	2	0	27
	Disagree	56	13	8	1	0	78
	Neutral	14	13	6	3	1	37
	Agree	7	23	11	2	0	43
	Strongly agree	1	7	5	2	0	15
Total		87	66	36	10	1	200
Can determine accurately the total debt position of my business	Strongly disagree	5	9	0	1	0	15
	Disagree	23	14	3	3	0	43
	Neutral	5	10	13	3	1	32
	Agree	41	33	20	3	0	97
	Strongly agree	13	0	0	0	0	13
Total		87	66	36	10	1	200

Source Primary data

Table 7.12 Budgeting literacy and growth in value of assets

		Value of assets					Total
		500 < 50,000	51,000 < 100,000	101,000 < 500,000	501,000 < 1,000,000	Over 1000,000	
Prepare annual budgets for the business	Never	60	29	3	0	1	93
	Rarely	10	12	3	3	0	28
	Sometimes	10	14	20	4	0	48
	Often	7	6	8	2	0	23
	Always	0	5	2	1	0	8
Total		87	66	36	10	1	200
Ability to compare financial objectives and business performance	Never	70	28	4	0	1	103
	Rarely	10	12	3	3	0	28
	Sometimes	7	14	20	4	0	45
	Often	0	6	8	2	0	16
	Always	0	6	1	1	0	8
Total		87	66	36	10	1	200
Use budgets to set targets for employees	Never	70	28	4	0	1	103
	Rarely	12	14	3	2	0	31
	Sometimes	5	12	20	6	0	43
	Often	0	6	3	1	0	10
	Always	0	6	6	1	0	13
Total		87	66	36	10	1	200
Use budgets to track spending	Never	53	14	3	0	1	71
	Rarely	17	17	6	1	0	41
	Sometimes	10	24	13	3	0	50

(continued)

Table 7.12 (continued)

		Value of assets					Total
		500 < 50,000	51,000 < 100,000	101,000 < 500,000	501,000 < 1,000,000	Over 1000,000	
	Often	5	4	9	5	0	23
	Always	2	7	5	1	0	15
Total		87	66	36	10	1	200
Prepare written annual financial objective	Never	50	28	3	0	1	82
	Rarely	20	12	3	5	0	40
	Sometimes	17	14	20	4	0	55
	Often	0	6	8	0	0	14
	Always	0	6	2	1	0	9
Total		87	66	36	10	1	200

Source Primary data

valuation 9 companies often use Budgets to Track Spending and for the 100,000/- company valuation 7 companies always use Budgets to Track Spending.

The results in Table 7.13 imply that there is a causal link between familiarity with financial services and an increase in the value of assets. 17 companies' valuation under 50,000/- No bank account is operated. 63 companies' valuation under 50,000/- said yes and their account is operated. 80 companies under the under 50,000/- said No for holidays. Whereas, 23 companies under the valuation of 100,000/- have said Yes to bank holidays for their business. For the assets under the valuation under 50,000/ - 15 companies said No to get extra finance for the business and 72 companies Yes. The companies under the 50,000/- and 100,000/- 61 companies do not Check visit the websites of commercial banks to learn more about the credit products they provide and the conditions for using them and 38 companies said yes check websites of commercial banks, in order to see the credit products, they provide and the conditions associated with them. For accessing goods available from the bank besides loans (credit-cards, fixed deposits, etc.). Under the valuation of 50,000/- of 83 companies said No and 66 companies under the valuation of 100,000/- have said Yes.

The results laid down in Table 7.14 elucidate the existence of a causal link between the ability to maintain accurate books and the rise in the value of assets. Value of Assets of below 50,000/-, 49 companies have No knowledge and Skills to Prepare Financial Statements. Ability to Perform Financial Analysis 51 companies has No knowledge and Skills. For the Maintenance of Cashbook 51 companies have No knowledge and Skills. For the Filling Tax Returns 59 companies have No knowledge and Skills and under the valuation of 50,000/-. For Maintaining and Balance the Ledger Accurately 57 companies have No knowledge and Skills and under the valuation of 50,000/- 58 companies under the valuation of 50,000/- have no knowledge and Skills Ability to do Bank Reconciliations.

Table 7.13 Banking services literacy and growth in value of assets

		Value of assets					Total
		500 < 50,000	51,000 < 100,000	101,000 < 500,000	501,000 < 1,000,000	Over 1000,000	
Operates a business bank account	No	17	3	1	0	0	21
	Yes	70	63	35	10	1	179
Total		87	66	36	10	1	200
Banks business proceeds daily	No	80	43	17	4	1	145
	Yes	7	23	19	6	0	55
Total		87	66	36	10	1	200
Knows where to get extra finance for the business	No	15	12	0	1	0	28
	Yes	72	54	36	9	1	172
Total		87	66	36	10	1	200
Is aware of commercial bank requirements of lending to MSEs	No	82	52	23	3	1	161
	Yes	5	14	13	7	0	39
Total		87	66	36	10	1	200
Checks commercial bank websites to see their credit products and their requirements	No	61	28	11	2	0	102
	Yes	26	38	25	8	1	98
Total		87	66	36	10	1	200
Can access other bank products other than loans (credit-cards, fixed deposits, etc.)	No	83	49	12	3	0	147
	Yes	4	17	24	7	1	53
Total		87	66	36	10	1	200

Source Primary data

Table 7.14 Book keeping literacy and growth in value of assets

		Value of assets					Total
		500 < 50,000	51,000 < 100,000	101,000 < 500,000	501,000 < 1,000,000	Over 1000,000	
Ability to prepare financial statements	No knowledge and skills	49	16	4	0	0	69
	Unsatisfactory knowledge and skills	24	19	4	1	0	48
	Satisfactory knowledge and skills	11	22	14	2	1	50
	Good knowledge and skills	2	7	14	6	0	29
	Excellent knowledge and skills	1	2	0	1	0	4
Total		87	66	36	10	1	200
Ability to perform financial analysis	No knowledge and skills	51	19	3	0	0	73
	Unsatisfactory knowledge and skills	25	20	6	4	0	55
	satisfactory knowledge and skills	9	19	13	2	1	44
	Good knowledge and skills	2	8	14	4	0	28

(continued)

Table 7.14 (continued)

		Value of assets					Total
		500 < 50,000	51,000 < 100,000	101,000 < 500,000	501,000 < 1,000,000	Over 1000,000	
Total		87	66	36	10	1	200
Maintenance of cashbook	No knowledge and skills	51	19	3	0	0	73
	Unsatisfactory knowledge and skills	24	17	4	1	0	46
	satisfactory knowledge and skills	10	23	14	3	1	51
	Good knowledge and skills	2	6	15	5	0	28
	Excellent knowledge and skills	0	1	0	1	0	2
Total		87	66	36	10	1	200
Filling tax returns	No knowledge and skills	59	25	4	1	0	89
	Unsatisfactory knowledge and skills	22	23	5	1	0	51
	satisfactory knowledge and skills	4	13	13	3	1	34

(continued)

Table 7.14 (continued)

		Value of assets					Total
		500 < 50,000	51,000 < 100,000	101,000 < 500,000	501,000 < 1,000,000	Over 1000,000	
	Good knowledge and skills	2	5	13	4	0	24
	Excellent knowledge and skills	0	0	1	1	0	2
Total		87	66	36	10	1	200
Maintain and balance the ledger accurately	No knowledge and skills	57	24	4	0	0	85
	Unsatisfactory knowledge and skills	23	23	4	2	0	52
	satisfactory knowledge and skills	5	12	14	2	1	34
	Good knowledge and skills	2	6	13	5	0	26

(continued)

Table 7.14 (continued)

		Value of assets					Total
		500 < 50,000	51,000 < 100,000	101,000 < 500,000	501,000 < 1,000,000	Over 1000,000	
	Excellent knowledge and skills	0	1	1	1	0	3
Total		87	66	36	10	1	200
Ability to do bank reconciliations	No knowledge and skills	58	24	4	1	0	87
	Unsatisfactory knowledge and skills	22	23	4	2	0	51
	satisfactory knowledge and skills	5	12	14	1	1	33
	Good knowledge and skills	2	6	12	5	0	25
	Excellent knowledge and skills	0	1	2	1	0	4
Total		87	66	36	10	1	200

Source Primary data

7.5 Conclusion

This research on microbusiness owners' financial literacy in Tamil Nadu's Coimbatore area produced some important results. First, it's important to address the problem of micro-entrepreneurs' poor financial literacy. According to the report, the majority of microbusiness owners are unaware of the numerous financial services, tools, and techniques available to them for managing their funds. Second, the research revealed that micro-entrepreneurs are more likely to depend on informal sources like friends and family than on official ones like banks and financial organisations for their financial information. The lack of access to legitimate sources and illiteracy in financial matters may be to blame for this dependence on unofficial sources. The survey also revealed that microbusiness owners are eager to contribute to financial literacy initiatives. This demonstrates that they are aware of the value of and potential advantages offered by financial literacy. As a result, the research comes to the key conclusion that financial literacy among micro-entrepreneurs in Tamil Nadu's Coimbatore area has to be addressed. Microbusiness owners that have superior those with a better understanding of finances are more inclined to make wise financial choices and have more success. The district's micro-entrepreneurs must be encouraged to become financially literate, and the government and other stakeholders must take action to do this.

References

Andarsari PR, Ningtyas MN (2019) The role of financial literacy on financial behavior. Jabe J Account Bus Educ 4(1):24. https://doi.org/10.26675/jabe.v4i1.8524

Andreou PC, Anyfantaki S (2021) Financial literacy and its influence on internet banking behavior. Eur Manag J 39(5):658–674. https://doi.org/10.1016/j.emj.2020.12.001

Ayhan B (2019) Constituting financialized subjectivities: cultural political economy of financial literacy in Turkey. Turkish Stud 20(5):680–707. https://doi.org/10.1080/14683849.2018.1520103

Development E (2018) rn all Sm of Bu es s a nd En te rp ris e D ev elo pm en rn all Sm of al nd En te rp ris e D pm. J Small Bus Enterp Dev 1–29

Devi SR (2020) Socioeconomic development of women through micro entreprises in Coimbatore District, no 09, pp 919–924

Gupta K, Kaur J (2014) A study of financial literacy among micro entrepreneurs in District Kangra. IMPACT Int J Res Bus Manag 2(2):63–70

Hasibuan BK, Lubis YM, HR WA (2018) Financial literacy and financial behavior as a measure of financial satisfaction, vol 46, no Ebic 2017, 503–507. https://doi.org/10.2991/ebic-17.2018.79

Kalaiselvi A, Krishnan K (2021) A study on customer's perception of evangelical social action forum in social banks (With Special Reference To Coimbatore District In Tamilnadu), p1711–1717. 7p

Kulathunga KMMCB, Ye J, Sharma S, Weerathunga PR (2020) How does technological and financial literacy influence SME performance: mediating role of ERM practices. Inf 11(6). https://doi.org/10.3390/INFO11060297

Mohd Thas Thaker MA, Amin MF, Mohd Thas Thaker H, Khaliq A, Allah Pitchay A (2021) Cash waqf model for micro enterprises' human capital development. ISRA Int J Islam Financ 13(1):66–83. https://doi.org/10.1108/ijif-08-2018-0091

Prasad H, Meghwal D, Dayama V (2018) Digital financial literacy: a study of households of Udaipur. J Bus Manag 5(I):23–32. https://doi.org/10.3126/jbm.v5i0.27385

Rachel Kuruvilla R, Harikumar PN (2020) A study on the financial awareness among women entrepreneurs in Kottayam district. J Manag Res Anal 5(3):331–335. https://doi.org/10.18231/2394-2770.2018.0052

Sangwan S, Nayak NC, Samanta D (2020) Loan repayment behavior among the clients of Indian microfinance institutions: a household-level investigation. J Hum Behav Soc Environ 30(4):474–497. https://doi.org/10.1080/10911359.2019.1699221

Shruthi K, Hiremath GM, Joshi AT, Patil SS (2018) An economic analysis of precision agriculture-a case study of paddy in North Eastern Karnataka. Indian J Econ Dev 14(2):274. https://doi.org/10.5958/2322-0430.2018.00130.0

Sibi MS (2021) Intelligent statistical modeling and approaches of financial literacy. https://doi.org/10.1109/ICACCS51430.2021.9441675

Singhal AK, Dhingra R, Pandey V (2015) The research repository. Anusandhan Res Repos 2(1)

Singla AA, Mallik G (2021) Determinants of financial literacy: empirical evidence from micro and small enterprises in India. Asia Pacific Manag Rev 26(4):248–255. https://doi.org/10.1016/j.apmrv.2021.03.001

Siregar (2022) No Titleםיניעה דגנל תמאבש המ תא תוארל השק יכה,” ץראה, vol 4632, no 8.5.2017, pp 2003–2005

Suganthi N, Asokhan M (2022) Marketing behaviour of registered women entrepreneurs in Coimbatore District of Tamil Nadu. Asian J Agric Extension Econ Sociol 40(10):1009–1014. https://doi.org/10.9734/ajaees/2022/v40i1031173

Suganthi N, Asokhan M (2022a) Perceived economic empowerment of registered women entrepreneurs in coimbatore district of Tamil Nadu. Curr J Appl Sci Technol 41(33):35–40. https://doi.org/10.9734/cjast/2022/v41i333949

Suthamathi D (2018) Research article a glass ceiling for first-generation women entrepreneurs in Coimbatore, Tamil Nadu, vol 4, no 1

Thomas B, Subhashree P (2020) Factors that influence the financial literacy among engineering students. Procedia Comput Sci 172(2019):480–487. https://doi.org/10.1016/j.procs.2020.05.161

Timcy S (2018) Financial literacy among people of Delhi: an empirical analysis. https://doi.org/10.5958/0973-9343.2018.00007.8

Tuffour JK, Amoako AA, Amartey EO (2022) Assessing the effect of financial literacy among managers on the performance of small-scale enterprises. Glob Bus Rev 23(5):1200–1217. https://doi.org/10.1177/0972150919899753

Ye J, Kulathunga KMMCB (2019) How does financial literacy promote sustainability in SMEs? A developing country perspective. Sustain 11(10):1–21. https://doi.org/10.3390/su11102990

Chapter 8
Technostress and Students in Sustainable Work Environment: An Empirical Study

K. Neelima Ravindran and M. Manjula

Abstract Technological advancements have significantly transformed the modern educational landscape, offering new opportunities and challenges for students. However, the widespread integration of technology in learning environments has also led to the emergence of a phenomenon known as "technostress" among students. This empirical study aims to investigate the impact of technostress on students in a sustainable work environment. The study employs a descriptive approach, using a quantitative survey method to gather comprehensive data from a diverse group of students. Participants were selected from various educational institutions and diverse socio-economic backgrounds to ensure a representative sample. The survey instrument measured the frequency and intensity of technostress experienced by students providing valuable insights into the underlying causes and coping mechanisms. Preliminary findings indicate that a significant number of students are affected by technostress in their educational pursuits.

Keywords Higher education · Students · Technostress

8.1 Introduction

With technological advancements and widespread use, ICT is being utilized more frequently in higher education institutions (HEIs). To enhance learning and automate work, technology is commonly employed in educational settings. In an effort to satisfy student expectations and gain from government incentives, technology-enhanced learning (TEL) has become increasingly adopted in academic contexts. Programs like integrated digital-based assessment, massive open online courses, learning management systems, student life cycle management, and attendance management systems are examples of how technology is being used for academic administration and student self-service. Student polls have shown that they are usually

K. Neelima Ravindran (✉) · M. Manjula
Dr. NGP Arts and Science College, Bharathiar University, Coimbatore, India
e-mail: neelimapmenon@gmail.com

A. Edward J. et al. (eds.), *Digital Transformation for Business Sustainability*, Contributions to Environmental Sciences & Innovative Business Technology,
https://doi.org/10.1007/978-981-99-7058-2_8

in favor of the use of ICT in the classroom, and online education and MOOCs have been found to cut the expenses of higher education for students. It has been shown that technology may assist HEIs in lowering bureaucracy and distance barriers.

ICT also makes it possible for HEI (Higher Education Institutions) to be transparent, process academic data more quickly, and streamline academic administration. The teaching and learning process is thought to be enhanced by the use of technology in the classroom. Although it is difficult to ignore the advantages of technology, there is a growing need to understand how it harms its users. In-depth analysis of technostress, or the "inability to change with changing technology," and its effects on work performance has been done on organizational people research. In a recent review, one of the six psychological and behavioral effects of technological stress was discussed, including how it negatively affects employee productivity.

There aren't many empirical researches that examine how prevalent technostress is among the younger population, especially among students. By causing students to be less productive, drop out, and deviate from their academic work, technology stress may increase the burden on higher education institutions. It is necessary to research how common technostress is among pupils as well as its effects. As they display a unique spectrum of traits and behaviors, today's students make for an intriguing group to study.

Due to the urgent use of online platforms in the midst of a scary epidemic, face-to-face instruction has been replaced, exposing personnel in higher education to new shifts. There have been some negative effects, notably on worker welfare and output. It is now important to comprehend how people have used technology to make remote employment possible. Every leader strives to achieve efficient academic and support staff operations under such circumstances. Understanding how the finest academic practices are evolving is vital for scholars. This study comprises six key sections. The introduction sets the stage, providing the context and significance of exploring technostress among students in sustainable work environments. The literature review probes into technostress manifestations in students, the role of technology in sustainability, and existing research in this intersection. The research methodology outlines the approach of study made, sample selection, data collection methods, and analysis techniques. Findings are presented, encompassing quantitative survey results on technostress. The discussion interprets the results, and exploring technostress implications for student well-being and academic performance. The conclusion summarizes key findings and reiterates the study's significance, ultimately emphasizing the importance of technostress management in fostering sustainable work environments for students along with recommendations, Limitations are acknowledged, and future research areas are proposed.

8.2 Literature Review

Scholars' curiosity on the negative effects of technology has grown over the last years. There is a wealth of material regarding professionals who experience technostress, or stress brought on by technology, in the literature. Although digital devices are more common in academic settings, there are few studies that address the frequency of technostress and its impacts among students. As a result of higher education's growing use of technology, students are now forced to use computers to complete all of their academic work, including exams. To utilize programs that employ technology to promote learning, such as learning management systems, MOOCs, and digital exam devices, students must develop ICT skills. The study examines the impact of technological stress on pupils' academic performance. The findings indicate that students had only mild levels of technological stress, and with just minor modifications, the technostress instrument may be used in a learning setting. Also, it was demonstrated that technological stress reduced students' academic performance (Upadhyaya and Vrinda 2020).

In this study, technostress among college students using technology-enhanced learning (TEL) was examined from a multidimensional person-environment mismatch viewpoint. Person-organization (P-O) mismatch, person-TEL (P-TEL), and person-people (P-O) misfit were the three aspects of technostress that were explored, in that order. A study model was created in order to analyze the relationships between the three dimensions of technostress and how they affected student burnout, persistence in TEL, and perceived performance. The findings show that P-O misfit has a strong predictive power for technostress on the P-TEL and P-P misfit dimensions. Technostress P-TEL mismatch was expected, as was Technostress P-P misfit. Their perceived burnout, which was closely associated with the three elements of technostress, had a substantial negative influence on the students' reported performance in TEL. Moreover, P-O technostress mismatch had the most substantial effect on students' burnout. Also, group comparisons based on gender and grade level reveal that females and students in lower grades were more susceptible than others to burnout associated with P-P mismatch of technostress. Also, research seemed that female students' performance was more negatively impacted by burnout than that of male students. The results of this study have important implications for pinpointing the elements affecting TEL students' academic development and welfare as well as for developing effective treatments for technostress (Wang et al. 2020a).

Technostress could be seen as a significant factor that could affect student performance and satisfaction. Very little research has been done on the simultaneous effects of the four components of technology stress—technological overload, technological complexity, technological insecurity, and technological uncertainty—on student satisfaction and performance expectations. Performance expectation may be used as an endogenous construct in this study because it was done during the open and distant learning (ODL) implementation and before the final exam. So, the aim of this study is to investigate the link between the four technostress features and student satisfaction. Also, this study attempts to look into the connections between undergraduate students' performance expectations and general student happiness. In this study, the measurement model was evaluated and the hypotheses were tested using SEM. This study found that technological complexity considerably adds more to student satisfaction and performance expectations than technological uncertainty. Yet, the results clearly indicate that there is no significant connection between students' satisfaction, performance expectations, and technology overload and insecurity (Abd Aziz et al. 2021).

University students are gaining in ways never previously imaginable as a consequence of technology-enhanced learning's growing prominence in higher education, but they may also experience technostress as a result of the greater demands and revised academic standards that come with it. The aim of this study was to develop a psychometric scale to measure the levels of technostress experienced by university students during technologically aided learning. Using the person-environment fit theory and related research, a prototype technostress scale was developed. In order to identify students who are not suitable for technology-enhanced learning and to protect their well-being in order to increase their positive and active participation in it, the development of the technostress scale is an essential first step (Wang et al. 2020b).

Booker et al. (2014), in higher education, information technology is being leveraged to provide students 24/7 access to resources. One of the factors influencing this acceptance is the students' evident demand for constant access to academic material. Technology-related stress, or technostress, is already common in today's modern environment. Working in an online learning environment increases this type of stress. This study provides a technique for calculating the stress that technological advancements cause among online students. Tools that are already used to assess general technology-related stress and technology-related stress in the workplace were looked at with the goal of establishing the initial study questions. The findings, which were studied using a convenience group of students, demonstrate a link between technological stress and academic performance in the online context.

Deng et al. (2022) examines the impact of academic and familial stress on students' depression levels and its subsequent effect on academic performance, drawing insights from Lazarus' cognitive appraisal theory of stress. Research consistently shows that academic stress leads to increased depression levels among students, negatively affecting their academic performance. Similarly, familial stress plays a significant role in students' mental well-being, either acting as a protective factor or exacerbating vulnerability to depression. Depressed students often experience difficulties in concentration, motivation, and cognitive functioning, resulting in poor academic outcomes.

8.3 Methodology

The survey method was employed to collect data from 94 B.com and M.com candidates at Dr. NGP Arts and Science College, specializing in International Business, Finance, Marketing, and Computer Applications. This empirical paper aims to analyze factors contributing to technostress among students. Percentage calculations were used for clarity and ease of analysis. The questionnaire captured technostress experiences, stressors related to technology use, coping mechanisms, and technology's impact on academic and personal lives. Sampling ensured representation from both B.com and M.com candidates across diverse specializations, enabling insights into potential variations in technostress. Anonymized responses ensured data accuracy and reliability. The use of percentages allowed straightforward presentation and comparisons between factors influencing technostress across various specializations.

8.4 Theoretical Framework

To start with a discussion on technostress the first thing enquired was if there was any vision problem for the students. Among the participants, a small portion of students had vision problems. Among those students suffering from vision problems, many students had vision problems only when using technological devices. Responses by the students on technology integration in their current study were discussed. As there were many learning management systems nowadays, these were compared on the basis of their ease of use. Many students responded as both online and offline mode of education is equally important. Technology inclusion in education was discussed and impacts were noted. Consequences of technology integration in education are discussed. When asked about online and offline mode of education, offline mode was preferred by the majority. Many of the students were feeling lonely when integrating technology into education. Technological inclusion helps students to think and work independently. Technologically driven classes are supposed to increase retention level of students. Many students feel PPT presentation method of teaching can reduce stress levels among students.

8.5 Analysis and Discussion

From Table 8.1 it is analyzed that among the participants, out of 94 student respondents 41.3% of them are male and 58.7% of them are female respondents. 37.8% of students was having vision problems. Among those students suffering from vision problems; 47.8% had long sight, 17.4% was having short sight and 34.8% of students had vision problems only when using technological devices.

Table 8.1 Demographic profile of respondents

	No. of respondents	Percentage
Gender		
Male	39	41.3
Female	55	58.7
Vision problem		
Yes	35	37.8
No	59	62.2
Type of vision problem		
Short sight	16	17.7
Long sight	45	47.8
Others	33	34.8

Source Author compilation

Table 8.2 depicts **students'** perspectives on technology integration in Education that 74.5% of the students considers technology integration in education as a good option. 76.6% of students considers there is above 50% technology integration their curriculum. Google classroom is considered as one of the best learning management systems by 70.2% of the students. 59.6% of the students consider both offline and online modes of education is equally important. Technology inclusion in education is considered as environment friendly and technostress increasing component by 31.9% of the students.

The Table 8.3 represents Students' Concerns and Issues with Technology Integration about 42.6% of the population considers technology integration may cause health issues while 38.3% believes it may cause psychological issues. 67.4% of the students are facing headache while using technical devices. Offline mode of education is preferred by 54.3% of students. 42.6% of students feel lonely on using technological

Table 8.2 Students' perspectives on technology integration in education

Perspectives	Number of students	Percentage of students (%)
Considers technology integration as good	70	74.50
Considers >50% technology integration in curriculum	72	76.60
Considers Google classroom as one of the best learning management systems	66	70.20
Considers both offline and online modes of education equally important	56	59.60
Considers technology inclusion as environment friendly and technostress increasing component	30	31.90

Source Author compilation

Table 8.3 Students' concerns and issues with technology integration

Concerns and issues	Number of students	Percentage of students (%)
Technology integration may cause health issues	40	42.60
Technology integration may cause psychological issues	36	38.30
Facing headache while using technical devices	63	67.40
Prefer offline mode of education	51	54.30
Feel lonely on using technological aids in education	40	42.60
Feel more stressed while attending technology-related classes	50	53.20
Technological inclusion helps students think and work independently	62	66.00
Technology-driven classes increase retention level of students	63	67.40
Believe resource persons are capable enough to deliver content with technological aids	60	63.80
Prefer PPT presentations over lecture methods, interactive seminars, and Google classrooms	43	45.70
Consider technostress as a physical state of mind	26	27.70

Source Author compilation

aids in their education. 53.2% feel more stressed while attending technology-related classes. 66% of the students agree technological inclusion helps students to think and work independently. 67.4% agrees that technology driven class's increases retention level of students. 63.8% of the students believe that resource persons are capable enough to deliver contents with technological aids. 45.7% of the students prefer PPT presentations over lecture methods, interactive seminars, and Google classrooms. Technostress is considered as a physical state of mind by 27.7% of the students.

8.6 Conclusion

Technostress being an important phenomenon nowadays, more and more studies should be made in this area to increase awareness among the same among people. Depression, loneliness, and burn out are some of the common concerns that need to be addressed. The study concludes with recommendations for educators, institutions, and policymakers to address the technostress phenomenon effectively and promote a sustainable work environment for students. Strategies such as digital detox initiatives, mindfulness training, and fostering a balanced approach to technology use are proposed to alleviate technostress and create a healthier learning atmosphere. By implementing these measures, educational stakeholders can pave the way for a more sustainable and student-friendly approach to technology integration in modern learning environments. Future study can be conducted on condition impacting both on physical and emotional state of mind of individuals, this need to be addressed efficiently to bring out best among students. Also, future research could involve longitudinal studies, cross-cultural comparisons, and objective measures of technostress. Addressing these aspects can lead to a more holistic understanding and effective strategies for managing technostress among students. Despite acknowledging limitations such as sample selection bias and self-reported data, the focus on a single institution may limit the generalizability of findings to other educational settings. However, the study's significance lies in shedding light on the technostress phenomenon and providing valuable insights for educators, policymakers, and institutions to develop effective strategies for managing technostress among students in sustainable work environments.

References

Abd Aziz NN, Awang Kader MAR, Ab Halim R (2021) The impact of technostress on student satisfaction and performance expectancy. Asian J Univ Educ 17(4):538. https://doi.org/10.24191/Ajue.V17i4.16466

Booker Q, Rebman CM, Kitchens FL (2014) A model for testing technostress in the online education environment: an exploratory study

Deng Y, Cherian J, Khan NUN, Kumari K, Sial MS, Comite U, Gavurova B, Popp J (2022) Family and academic stress and their impact on students' depression level and academic performance. Front Psychiatry 13:869337. https://doi.org/10.3389/fpsyt.2022.869337

Upadhyaya P, Vrinda (2020) Impact of technostress on academic productivity of university students. Educ Inf Technol 26(2):1647–1664. https://doi.org/10.1007/S10639-020-10319-9

Wang X, Tan SC, Li L (2020a) Technostress in university students' technology-enhanced learning: an investigation from multidimensional person-environment misfit. Comput Hum Behav 105:106208. https://doi.org/10.1016/J.Chb.2019.106208

Wang X, Tan SC, Li L (2020b) Measuring university students' technostress in technology-enhanced learning: scale development and validation. Australas J Educ Technol 96–112. https://doi.org/10.14742/Ajet.5329

Chapter 9
Indian Economic Crisis: COVID-19 Versus Russia-Ukrainian War

Salma Begum and R. Guna Durga

Abstract The economic stability of a country is important for the country's economic growth. This paper focuses on the economic crisis India has been facing currently due to inflation. The study is based on the impact of COVID-19 and the Russia-Ukrainian on the Indian economy. The objective of the paper is to analyse the economic crisis of India and the policy measures taken by the government. The current study is based on an extensive literature review and secondary data analysis. The secondary data has been used from different published reports. Trend analysis is done for unemployment and inflation to measure the impact of the crisis on the Indian economy. It was found that India was facing a range of economic issues that have both short-term and long-term implications. The country needs to take proactive steps to address the issues and improve the economic outlook of the nation.

Keywords Indian economic crisis · Inflation · Russia-Ukrainian war · Oil prices

9.1 Introduction

The current economic crisis in India is a result of a combination of factors, including the COVID-19 pandemic, the oil price crash and a slowdown in global demand. It has caused a severe slump in the Indian economy, with GDP growth rates falling to their lowest levels in decades. The oil price crash has also had an adverse impact on India's economy, as the country is heavily dependent on oil imports for its energy needs. The sharp fall in the price of oil has resulted in a significant loss of revenue for the Indian government, with the budget deficit rising to 6.8% of GDP in 2020–21. Finally, the slowdown in global demand has also had a negative effect on India's economy, as the country is heavily dependent on exports for its economic growth. Exports have declined significantly since the onset of the economic crisis, with exports of

S. Begum (✉) · R. Guna Durga
CMS Business School, Faculty of Management Studies, JAIN (Deemed-to-be University), Bangalore, Karnataka, India
e-mail: salma.begum999@gmail.com

A. Edward J. et al. (eds.), *Digital Transformation for Business Sustainability*, Contributions to Environmental Sciences & Innovative Business Technology,
https://doi.org/10.1007/978-981-99-7058-2_9

goods and services falling by 23% in 2020–21. The decline in global demand has had a negative impact on India's manufacturing sector, further reducing economic activity. The Indian government has implemented several measures to address the crisis, including a massive stimulus package, an expansion of the MSME sector and an increased focus on digital infrastructure (Bhattacharya and Patnaik 2014). Despite these efforts, the crisis has continued to deepen, leading to a sharp contraction in GDP growth, rising unemployment and a widening of the country's fiscal deficit. It is clear that the current economic crisis in India is one of the most severe in decades, and its resolution will require long-term economic reforms and an integrated approach to policymaking.

The following sections consist of a literature review to unearth the research gap. The following section highlights the economic crisis and the policies that were taken to tackle the crisis. Finally, the conclusion section gives the suggestions to the pertinent problem along with the study's limitations and the future scope for further research.

9.2 Literature Review

Ghosh and Chandrasekhar (2009) disprove the myth that, due to its low level of integration and other natural advantages, the Indian economy will not be impacted by any type of global financial crisis. In reality, however, India's prior dependence on export, capital flow, and supporting credit-fueled consumption and investment will depend greatly on global integration. Both the domestic and global economy will be impacted by the growing process.

Walia (2012) opines how the world economy is actually having a detrimental impact on the rapidly expanding Indian economy. The Indian economy claimed during the global financial crisis of 2001–2002 that it was unaffected by the US housing market disaster, but later came to admit that there was some impact. The paper argues that various economic crises have both beneficial and negative effects on the Indian economy.

Kumar (2009) found that the Indian economy has been severely impacted by the effects of the global economic crisis in three key areas: finance, commerce and exchange rates. Although the Indian economy has not been seriously impacted by the micro-crisis, just 22% of GDP is contributed by exports of goods and services. In 2009, India saw a dip in the GDP growth rate due to a drop in exports. By improving the investment climate, lowering entry barriers for corporate investment, upgrading the provision of public goods and services, and boosting the physical infrastructure's capacity, fiscal policy should remove structural obstacles that prevent a fall in GDP growth.

Das et al. (2012) explain that US economic crisis that occurred in 2008–2009 as a result of sub-prime mortgages, which led to homeowners making excessive mortgage payments and facing foreclosure. Due to weak global corporate governance in the US, excessive mortgage payments caused an imbalance between financial innovation and regulation. Developed countries suffered a decline in FDI inflow, which had the biggest impact on their economies as a result of the crisis. The Indian economy and agriculture are not completely immune to domestic and global economic downturns.

Dev and Sengupta (2020) observed that even before the COVID-19 outbreak, the Indian economy had already suffered greatly; the shutdown just made things worse. The study describes the pre-COVID 19 economic situation in India, assesses the policies adopted by the Indian government and discusses how India managed to emerge from this crisis.

Mehta and Jha (2020) undertook a thorough analysis of the macroeconomic effects of numerous health crises that have already happened and projected the impact of the COVID-19 pandemic in India. All countries' priorities during the COVID-19 pandemic were the slowing of the virus's spread rather than the economy, as this minimized the effects of the global slump. The research advocates building a framework for a macroeconomic model that accounts for the relative costs and effects of an outbreak on a global scale.

Mohammed (2022) demonstrated a number of economies' inflation rates that have been impacted by the conflict between Russia and Ukraine, which contributed to the global economic crisis. The price of various products has increased in the countries due to the war between Russia and Ukraine, which is the producer of significant exports. Petroleum has had tremendous impact on the current account balance, particularly on the Indian economy. As a result, consumers are suffering.

Singh and Verma (2016) ran a regression analysis on GDP, unemployment and inflation; the outcome demonstrates that there is an insignificant relationship between unemployment on real GDP and Inflation. The paper examines the trade-offs between India's inflation and unemployment from 2009 to 2015. According to the report, policymakers should concentrate on the level of market volatility, the employment rate and the restructuring of the Indian economy.

Veni and Choudhury (2007) stated that growth and inflation are subjected to a causality test in the study, which conclusively demonstrates that they are two distinct factors. A co-integration test is also undertaken, which demonstrates that the two variables are not associated. The test period included the years 1981 to 2004. As a result, the analysis indisputably shows that there is no long-term relationship between inflation and economic growth in India. According to the study, the Indian government should promote economic expansion and take prompt action to lower inflation.

The above studies confirmed that there were no studies on the Indian economic crises based on trend analysis.

9.3 Research Methodology

The current study is based on an extensive literature review and secondary data analysis. The secondary data has been used from different published reports. Trend analysis is done for unemployment and inflation to measure the impact.

9.4 Economic Crisis in India

Over the last two years, there have been too many supply chain disruptions caused by COVID-19 and the Russia-Ukrainian war. During COVID-19 due to the shortage of supply, the government and central bank injected excess liquidity to beat the recession but that led to demand-pull inflation. The COVID-19 outbreak in India has sparked an economic collapse and a rise in pricing. This has been driven by a combination of factors, including the disruption of supply chains, reduced consumer spending and increased costs of production. The reduced purchasing power of consumers due to job losses and pay cuts has further contributed to rising inflation. The WPI-based food inflation was 6.8% in November 2021 and increased to 9.93% in August 2022.

Figure 9.1 shows that the unemployment rate in India was 7.3% before the onset of the pandemic in July 2019 and then gradually increased to 8.8% in March 2020. After that, the unemployment rate decreased to 7.4% in July 2020 and then further decreased to 6.5% in November 2020. The inflation rate in India was 3.1% in July 2019 and then gradually increased to 5.8% in March 2020. After that, the inflation rate decreased to 6.7% in July 2020 and then further decreased to 5.5% in November 2020. Overall, the data indicates that the COVID-19 pandemic has had a significant impact on the economy of India. The pandemic has caused a rise in unemployment rate and inflation rate in India, but the effects have been gradually decreasing over time. The demand-pull inflation has been caused by an increase in the demand for essential goods, such as food and medical supplies, due to the rise in the death rate. This has led to an increase in the prices of these goods due to the limited supply. Additionally, the government's decision to provide relief packages to those affected by the pandemic has also increased demand, leading to higher prices.

Further, there has been increase in demand for basic necessities in Indian Households because of the socio-economic condition. With limited purchasing power, households have had to spend more on essential goods and services, leading to a further rise in prices. Demand-pull inflation is a type of inflation caused by an increase in aggregate demand in the economy. During the COVID-19 pandemic in India, the socio-economic status of many Indian households has caused an increase in demand for basic necessities such as food, housing and healthcare. As the demand for these basic necessities increases, their prices also rise, resulting in demand-pull inflation. In addition, as the pandemic has led to a decrease in the incomes and employment opportunities of many people, they are forced to resort to borrowing and spending, which further adds to demand-pull inflation. As a result, the cost of living is increasing, leading to a decline in the purchasing power of Indian households.

The economy was working towards revamping all the sectors but Russia-Ukrainian War led to the worsening of the crisis situation. The war has had a major impact on the global economy, including India. The conflict has resulted in higher prices for many commodities due to increased demand from the warring countries and a decrease in the availability of these commodities. This has caused India to suffer from cost-push inflation, which is when prices increase due to an increase in the cost of production. The WPI spiked to 15.88% and retail inflation at 7.3%

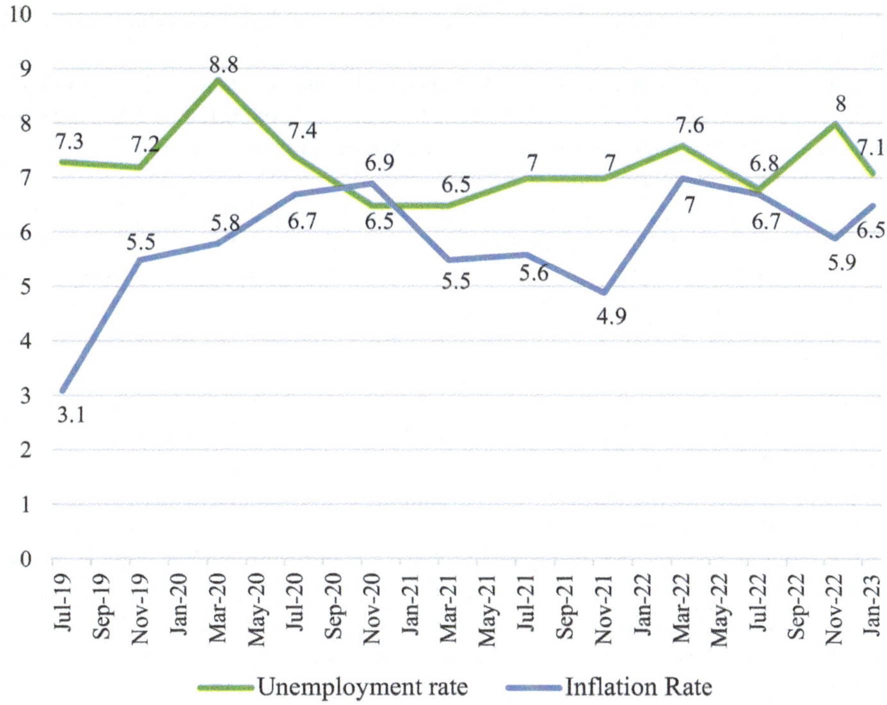

Fig. 9.1 Trend analysis: unemployment rate versus inflation rate

in September 2022. This is due to the rise in prices of minerals, oils, food articles, crude oil, etc. High food price inflation threatens the living standard of the majority of our population as food expenses amount to more than 50% of average household spending.

Russia and Ukraine are significant sunflower oil producers due to the volatility of various foreign governments that have taken protectionism measures. Indonesia banned palm oil exports on May 2022 to reduce export levies and taxes due to excessive domestic supplies. India in response to that issued a zero-duty tariff quota of 4 million metric tonnes of soybean oil and sunflower oil alleviating inflation. Also, India has imposed a ban on the export of certain food items such as wheat, maize, pulses and cooking oils. This ban was put in place to ensure that the domestic food supply remains secure and to prevent further price increases. In order to satisfy the rising demand, India has also taken initiatives to improve food production.

As is evident from Fig. 9.2, total export has been rising steadily both before and after the pandemic, but exports of agricultural products have been declining since 2022. Until that point, exports of agricultural products were rising, and even during the pandemic period, agriculture was the only industry to demonstrate growth, particularly in terms of commerce. However, due to the Russia-Ukraine war, exports of agricultural products have been purposefully declining to lessen reliance on food imports.

The shortage of food caused by the Russia-Ukraine war also had a significant impact on India's export sector. Many of the food products exported by India are now in short supply, which has led to a decrease in exports. This has caused India to suffer from a decrease in international trade and a loss of income, resulting in a negative impact on the country's economy. Oil production was reduced by 2 million barrels per day (bpd) on October 4 by the Organization of Petroleum Exporting Countries (OPEC) and its partners, known as OPEC+, which helped crude oil futures. Beginning in November, OPEC+ reduced oil output by 2 million bpd, citing the "uncertainty that surrounds the global economy and oil market outlooks" as the reason for the move. Due to sanctions against OPEC countries like Russia, Venezuela and Iran as well as production concerns in Nigeria and Angola, the actual decline would be between 1.0 and 1.1 million bpd. When OPEC cuts its production, it reduces the supply of oil and this leads to a rise in oil prices. This is because a reduction in the supply of oil reduces the amount of oil available for consumption, and the resulting scarcity causes an increase in price.

For India, an increase in oil prices will have a significant impact. India is one of the largest importers of oil in the world and relies heavily on crude oil imports to meet its energy needs. As such, an increase in oil prices will lead to a significant increase in India's import bill, putting a strain on the country's economy. Additionally, it will increase inflation and reduce the purchasing power of consumers, thus making it

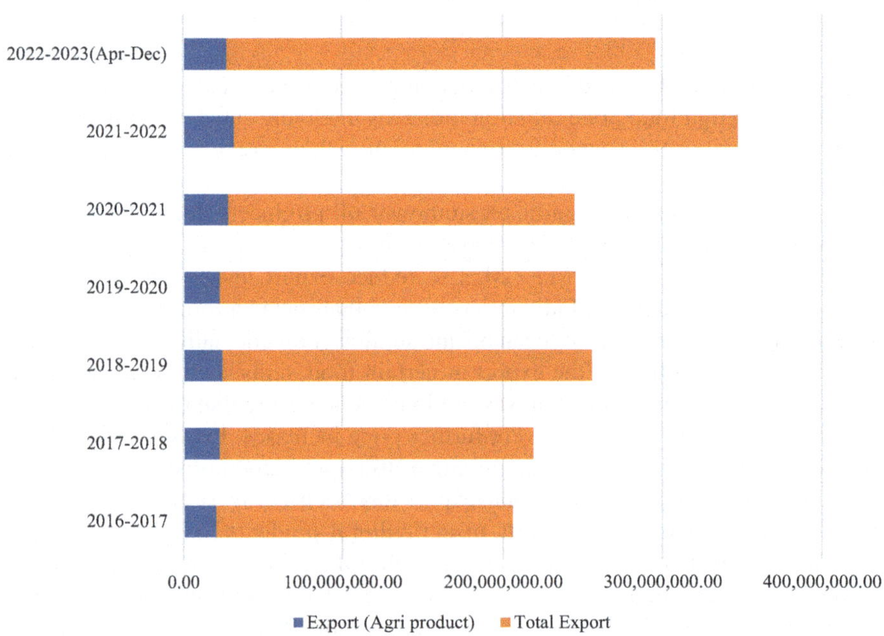

Fig. 9.2 Indian export from 2016 to 2023

more difficult for people to afford basic necessities. Finally, higher oil prices may also lead to a decrease in investment as companies struggle to cover the higher cost of oil. To solve this problem of petrol and oil consumption, India has introduced many schemes and tax concessions under Budget 2023 to promote E-vehicles among Indian users like imposing Customs Duty reduction from 21 to 13% on Lithium Batteries, which reduces the cost of producing the E-vehicles and further increase the sale of E-vehicle.

9.5 Government Policies to Check Economic Crisis

The Indian Government has implemented a number of policies to control inflation during the COVID-19 pandemic. These policies aim to reduce the prices of essential commodities and to provide financial aid to those affected by the crisis. One of the main policies is the Pradhan Mantri Garib Kalyan Yojana (PMGKY), which provides direct cash transfers to vulnerable individuals and families. The government has also set up a special fund to provide direct benefit transfers to the most vulnerable households in the country.

- The government has also reduced import duties on essential commodities such as pulses, edible oils and other food items. This has helped to reduce the prices of these items, making them more affordable for consumers.
- Import restriction on selected pulses has been removed on 15 May 2022 as the government allowed unrestricted import of black gram lentils, mung beans, etc. till 2026.
- The government has also implemented a number of measures to reduce the cost of production and distribution of essential commodities, such as reducing the cost of freight and transportation, waiving or reducing the cost of storage and providing tax benefits to producers.
- The government has also introduced an Interest Subvention Scheme for small and medium enterprises to help them sustain their operations during the pandemic. Additionally, the government has set up a special fund to provide credit to MSMEs. These policies have helped to reduce the prices of essential commodities and to provide financial assistance to those affected by the crisis. This has helped to control inflation in the country.
- India has been diversifying sources to reduce oil dependency on a few countries. Indian oil corporation recently linked a long-term supply deal with Brazil's Petrobras and Columbia's Ecopetrol.

9.6 Conclusion

In conclusion, India is facing a range of economic issues that have both short-term and long-term implications. The country needs to take proactive steps to address the issues and improve the economic outlook of the nation. This can be done through policy reforms, better utilization of resources and making sure that the benefits of economic growth are shared with all sections of society. India also needs to focus on creating jobs, improving access to capital and increasing the efficiency of the financial sector. With the right policies and investments, India can become a global leader in the economy and provide a better quality of life for its people. Even India came up with an increase in the Repo rate for the 6th time and the recent budget is also helping to control inflation. For instance, India came with many skill-based courses for youngsters to decrease the unemployment rate in India. Government has to ensure that there is a balance between the fiscal situation and the income support, therefore the government should find the maximum opportunity for poor people for earning, which depends on the hands of municipal and state governments. The paper has taken a timeline of events that took place only between 2016 and December 2023. It took into consideration only about the export, unemployment and inflation rate into consideration. Other factors like foreign exchange rate, balance of payment, Gross National Product could be taken in the future research.

References

Bhattacharya R, Patnaik I (2014) Monetary policy analysis in an inflation targeting framework in emerging economics: the case of India

Das A, Kumar NR, Debnath B, Mandal SC (2012) Global economic crisis: causes, impact on Indian economy, agriculture and fisheries. Int J Agric Sci 4(4):221–226

Dev SM, Sengupta R (2020) COVID-19: impact on the Indian economy. Indira Gandhi Institute of Development Research, Mumbai

Ghosh J, Chandrasekhar CP (2009) The costs of 'coupling': the global crisis and the Indian economy. Camb J Econ 33(4):725–739

Kumar R (2009) Global financial and economic crisis: impact on India and policy response. United Nations Development Programme

Mehta K, Jha SS (2020) COVID-19: a nightmare for the Indian economy. UGC Care J 31(20)

Mohammed N (2022) The Russia-Ukraine war crisis—it's impact on Indian economy. Int J Res Anal Rev (IJRAR)

Singh D, Verma N (2016) Trade-off between inflation and unemployment in the short run: a case of the Indian economy. Int Financ Bank 3(1):77

Veni LK, Choudhury PK (2007) Inflation and growth dilemma: an econometric analysis of the Indian economy. IUP J Financ Econ 5(1):79–87

Walia S (2012) Impact of global economic crisis on Indian economy: an analysis. Int J Latest Trends Eng Technol 1(2):31–36

Chapter 10
Digital Transformation in Sustainable Digital Marketing: Trends, Opportunities and Challenges in Targeting Millennial Consumers

Ganesh Hosur and Kattamuri Satish

Abstract The complexity of consumer behaviour challenges marketers to keep up with the rapidly proliferating digital channels and their users. The most targeted segment of digital marketing is 65% of India's population, the young consumers. The consumer market is dominated by millennials, born between 1980 and 2000, who are young, learned, affordable, affluent, tech-savvy and decision-makers when it comes to purchases. They are the biggest part of economic growth and development because of their unique characteristics. This study intends to measure the impact of digital marketing channels on the millennials. It was found that the advertisement needs to be crafted specially for this segment. Unique colour, feel and content are needed for reaching out to this section of the society.

Keywords Digital transformation · Sustainable digital marketing · Millennial consumers

10.1 Introduction

Digital transformation is an integral part of every business, organization, industry and academic institution including the Government and its various departments. Cutting-edge digital technologies like the Internet of Things, Robotics and Robotic Process Automation, Augmented Reality, Cloud Technology, API integrations, Artificial Intelligence and Machine Learning, Big Data, Real Time Analytics and Mobile Technology are some of the tools being used in manufacturing, marketing and distribution of the marketing communications, products and services. Integration of these technologies is not only making companies competitive but also providing value to

G. Hosur (✉) · K. Satish
CMR University, Bangalore, India
e-mail: ganesh.hosur@cmr.edu.in

A. Edward J. et al. (eds.), *Digital Transformation for Business Sustainability*, Contributions to Environmental Sciences & Innovative Business Technology,
https://doi.org/10.1007/978-981-99-7058-2_10

their customers and other stakeholders. Challenges encapsulated with digital transformation are data privacy, data security, digital literacy and fueling inequality between digitally connected and not connected populations. The road ahead for digital transformation has unlimited opportunities, despite a few challenges to overcome.

Rapid changes are happening in Digital Marketing due to the digital transformation in all the connected fields. Digital marketing uses various digital media and channels to communicate marketing messages to the audience and to supply products and services to the customers. Through marketing activities, companies promote their products and services to increase brand awareness and improve sales and market share. Digital marketing strategies are becoming more powerful as well as dynamic to reach a larger audience through social media platforms. Companies are striving very hard to understand the constantly changing consumer behaviour. AI tools and market research are extensively being used to understand the consumer thought process and purchase behaviour. Inputs from the descriptive, diagnostic and predictive analysis are helping the companies to effectively use the enormous data being tracked, to develop meaningful and impactful marketing communications. Some of the channels used for digital marketing are Website Marketing, Pay-Per-Click Advertising, Content Marketing, Email Marketing, Social Media Marketing, Affiliate Marketing, Video Marketing, SMS Messaging, Search Engine Optimization and Search Engine Marketing. The use of these digital marketing channels is making sure that the marketing communications are reaching a larger consumer audience, globally.

Consumer characteristics of millennials are very unique as they are born in the digital era and are always connected to the internet. They are family-oriented, well-informed, well-connected and fun-loving. They make informed and responsible buying decisions. Green marketing, green consumerism and environmental sustainability are the latest trends embraced by millennials because of their consciousness and concern for the environment and society. Millennials are impressed by artistic, creative, quirky, fun and offbeat humorous advertisements which are created with stunning visuals. Millennials' decision-making process is also complex as it depends upon many factors like age, family size, disposable income, lifestyle and the influencer group like celebrities, social media influencers and the peer group.

This study will bring out the latest trends, opportunities and challenges by reviewing earlier studies on digital transformation, digital marketing, marketing strategy and millennial consumer characteristics. The study will also highlight the findings and recommendations, based on the outcome of earlier studies, which can be considered by digital marketers to change their marketing strategies and marketing communications to connect with the target audience in the future.

10.2 Objectives of the Study

The primary objective of the study is

(a) To study the impact of digital transformation on digital marketing and marketing communications, targeting millennial consumers.

The sub-objectives of the study are

(a) To understand the components of digital transformation and digital marketing.
(b) To bring out various characteristics of millennial consumers in digital marketing.
(c) To study the digital marketing strategies.

10.3 Literature Review

Following review/research studies that are relevant to the study topic and the objectives of the study are chosen.

Digital transformation is the hottest research topic as it has brought a great deal of change in society, the economy and industry. Digital technology and systems are revolutionizing digital development which has resulted in innovation practices, improved designs, and new business models like Airbnb and Amazon.

Dash and Chakraborty (2021) found that search engine marketing and search engine optimization, display and E-CRM practices impacted consumer satisfaction and also their purchase intentions. The study also revealed that consumer satisfaction influences purchase intention and it also played a mediator role between digital marketing practices and purchase intention.

A study by Soto-Acosta (2020) found that the COVID-19 pandemic has shifted the digital transformation to high-speed gear. The study result found that digital transformation compelled all industries to be more competitive and relevant. As per the author, organizations are accelerating the adoption of digital transformation as the best way to avoid short-term economic collapse and to combat any future unexpected situations, with resilience.

Reinartz (2018) study found that the rise of e-commerce, mobile shopping and smart technologies is creating new competitors in the digital marketing space. The author suggests that businesses have to acknowledge the rapid changes happening through digital transformation and proactively manage their competitive position in the continuously evolving ecosystem.

Spilotro (2016) studied the Internet of Things revolution in the digital marketing landscape in influencing millennials. As per the author, the growing phenomenon of the IoT provides an unprecedented opportunity for businesses to target millennial consumers. The study reveals that millennial generations are quick to adapt to IoT and their growing purchasing power makes them the ideal target market.

Moreno (2017) studied the characteristics of millennials and their purchase behaviour. The results revealed that the millennials are a highly attractive market and this generation is in an environment where technology provides a platform for immediate gratification and personalization. The buying process for millennials is more fun, whereas loyalty to the brand is relative. The study results show that coupons, discounts, cashback and points as rewards, stimulate millennials toward purchase intention.

Melovi (2021) studied the determinants of the purchase behaviour of millennials in digital media shopping and its implications on customer satisfaction. The results revealed that millennials are satisfied with their online purchases though they perceive some risk in it. The results showed that there is no gender difference concerning the frequency of online purchases, though men spend more money than women. Younger millennials purchase more often than older millennials and men buy more from popular websites than women do.

Tan (2019) studied the factors which influence green product purchase behaviour among millennials. The study results indicated that environmental consciousness, eco-label and advertising were important predictors of green product purchase behaviour among millennials.

Munsch (2021) studied digital communication and advertising effectiveness on millennials and generation Z. The study result found that marketers are being challenged to uncover new insights to hold millennial consumer attention, as a digital distraction is a common behaviour during an advertisement. The study results revealed that digital marketing communications that are short, musical, humorous, engaging, informative and promoted by social media influencers have a positive impact on both age cohorts.

Dunlop (2016) studied the marketing strategies aimed at youth in the digital age. The study reveals that huge advertising budgets of global commercial brands allow them to experiment with new and exciting ways to entice youth to think about, share information and try products that are detrimental to their health.

Naumovska (2017) studied millennials' content preferences and media preferences for marketing communication campaigns. The study found that millennials prefer social and digital media rather than traditional media when getting information about brands, products and services. Marketing communications with graphics and strong visuals, overwhelmingly attract the attention of Millennials. The study also found that honesty and fun are the most preferred marketing communication content, expected and valued by the millennials.

Smith (2011) studied the various types of digital media marketing strategies of companies, to ascertain the advertising communications which are found to be appealing, motivating, or just annoying to the millennials and which ones are influencing their purchase behaviour. The study result revealed that millennials prefer online coupons and side-panel advertisements but don't like pop-up advertising. Graphics and visuals with bright colours and interactive features are more effective in grabbing their attention, while competitive prices and shipping rates will influence millennials to visit a website repeatedly.

Gupta (2021) studied young consumers' attitudes toward digital media advertising, on Instagram, which is considered to be one of the most popular and influential social media platforms followed by the youth, consistently. The study found that informative content plays a significant role in developing positive attitudes toward advertising. Humour-laced entertaining content emerged as an important predictor in the study.

The above studies pointed to the fact that there have been no prior studies conceptually evaluating the impact of digital marketing on millennials.

10.4 Discussion

Digital transformation has impacted all aspects of the economy and businesses, from production to consumption, including marketing and distribution. Personal computers, laptops, tablets and mobile devices, which are seamlessly connected to the faster internet, have changed the way marketers are reaching, engaging and interacting with consumers around the globe. The magnitude of the technological developments and adoption of the past five years has tremendously changed India's image in the world as the fastest-growing economy. Digital transformation combined with digital technologies is bringing the nations closer for collaboration, technology transfer, knowledge sharing and growth partnership. Business enterprises have to acknowledge the rapid changes happening through digital transformation and proactively manage their competitive position in the constantly evolving ecosystem.

Figure 10.1 shows digital technology tools as components of digital transformation which are used in processes and strategies to create competitive market growth, market share, customer value and continuous engagement with all the stakeholders.

Digital transformation has led to extensive use of digital media to market messages, products and services to consumers, irrespective of their geographical location. Digital marketing through digital media is growing at a phenomenal pace, to reach global consumers, with innovative marketing communications on products and services. Technologies are constantly evolving to bring marketers and consumers closer to each other with more engagement and interaction. Consumers are spending more time on the internet hence marketers are focusing on digital media and digital marketing techniques to effectively attract, engage and delight online audiences.

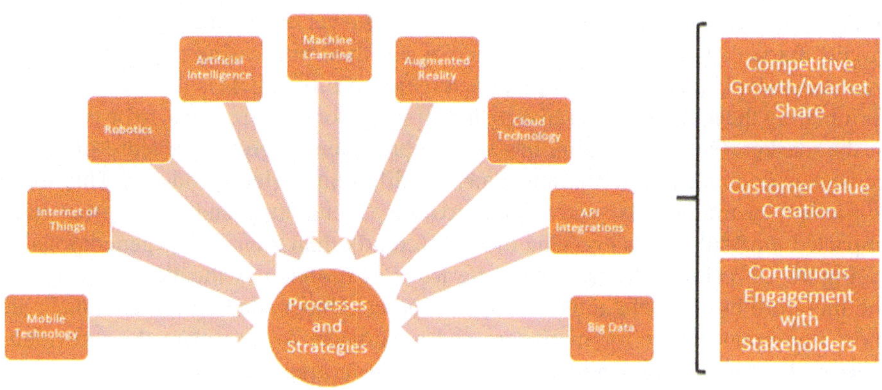

Fig. 10.1 Digital transformation components. *Source* Author compilation

Fig. 10.2 Digital marketing components

Digital marketing uses mobile devices, websites, search engines and social media to connect, engage, market and interact with consumers. The biggest challenge faced by marketers, in the digital world, is how to set themselves apart amongst the overcrowded digital marketing advertisements. The success of digital marketing campaigns depends upon a deeper understanding of consumer behaviour, which includes customer preferences and motivations.

Figure 10.2 shows digital marketing components which are focused on consumers to deliver marketing communications, products and services.

Millennials are the generation born between 1980 and 2000 and they are currently between 23 and 43 years. They are the market drivers and trendsetters of economic development and growth. All the organizations in the industry are vying for a slice of the market share of the millennials' purchase power. Millennials are not only the largest market segment; it is also attractive and their purchasing power is growing tremendously. Understanding the characteristics of millennials is the foundation for formulating the marketing and advertising strategy, for organizations, to be the leader in the market. Online shopping has seen unprecedented growth because of strong consumer demand and increasing types of different variety products. This trend is fuelled by millennials' active participation. Technology and the internet have impacted the Millennials evolving them differently than their earlier generations.

Figure 10.3 shows different characteristics of millennials identified in some of the literature by multiple authors in their studies. Understanding millennials is the most important aspect for marketers and advertisers to formulate the most suitable marketing communication content while promoting their brands, products and services. The idea is to impact their purchase intentions and behaviour, which may turn into brand loyalty in the long term.

Fig. 10.3 Millennial consumers

Millennials are the largest but distinctive market segment in demographic and psychographic characteristics, whose thoughts, feelings, behaviours, attitudes, habits and buying behaviour needs to be understood by marketers and advertisers to tap the full potential of their purchase power. Millennials are highly influenced by technology, digital media, and the internet, and the population size is large, hence targeting them with attractive marketing campaigns and sales promotion seems to be a challenging task for marketers and brands. Marketers and advertisers consider consumer-to-consumer promotion in viral advertising, as an opportunity to create awareness, attract attention and imbibe trust in their brands while remaining in control of the format and content of the marketing communication. Psychological motivational factors are also being considered for marketing communication content strategy. Millennials are the driving force and trendsetters of online shopping. This growth in

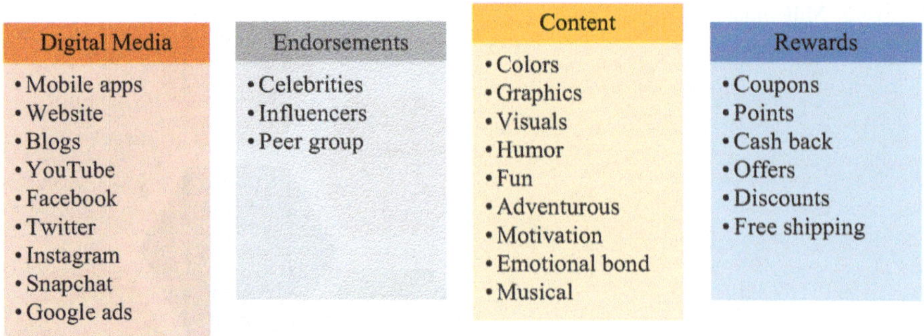

Fig. 10.4 Marketing strategy

digital media usage is forcing more companies to devise innovative digital marketing strategies to target potential consumers. In this direction, companies are attempting to create brand love through emotional bonds with consumers, which is anticipated to influence commitment and positive word-of-mouth promotion.

Figure 10.4 shows the approach of marketers and advertisers in attracting the attention of millennials. Millennials are immersed in digital media hence marketing communications are most commonly targeted through these media channels. Advertisements use colours, graphics and catchy phrases to connect with millennials emotionally and influence their feelings and actions toward purchasing their products and services. Using young and popular celebrity and influencer endorsement along with rewards like coupons and cashback are the bigger steps in reaching out to millennial consumers.

10.5 Conclusion

Digital transformation has changed the way people learn, work, communicate and collaborate. Technological inventions are making a seamlessly connected world into a gigantic information system that produces enormous data to analyse, interpret and use. Digital transformation, through digital technologies, has brought massive changes in strategies, processes and cultures of organizations and industries to be relevant in the competitive world. Digital transformations, Digital Media and Digital Marketing are all evolving and changing as and when new technologies are invented or adapted. The industries and leadership have to be on the edge to dynamically change towards growth and progress, whenever new inventions disrupt the existing ecosystem.

Near-ubiquitous use of social media by young adults creates enormous opportunities for marketers and advertisers to use digital media to promote their brands, products and services, with attractive marketing communication content.

Digital marketing advertisements which are short, crisp and humorous, with stunning visuals and endorsement of celebrity and social media influencers, can have a big impact on younger generations to attract and influence their behaviour and purchase decisions.

Millennials are the largest but distinctive market segment in demographic and psychographic characteristics, whose thoughts, feelings, behaviours, attitudes, habits and buying behaviour needs to be understood by marketers and advertisers to tap the full potential of their purchase power.

Green consumerism and sustainability are the latest trends embraced by the millennials, because of the heightened awareness of the sustainable environment and social concern in preserving the green environment for future generations.

References

Dash G, Chakraborty D (2021) Digital transformation of marketing. Sustainability. https://doi.org/10.3390/su13126735

Dunlop S (2016) Marketing to youth in the digital age: the promotion of unhealthy products and health promoting behaviours on social media. Media Commun. https://doi.org/10.17645/mac.v4i3.522

Gupta M (2021) Instagram advertising among young consumers in wearable fitness trackers: the moderating role of technology acceptance factors. J Glob Mark. https://doi.org/10.1080/08911762.2021.1931616

Melovi B (2021) Determinants of millennials' behavior in online shopping—implications on consumers' satisfaction and e-business development. Technol Soc. https://doi.org/10.1016/j.techsoc.2021.101561

Moreno FM (2017) The characterization of the millennials and their buying behavior. Int J Mark Stud. https://doi.org/10.5539/ijms.v9n5p135

Munsch A (2021) Millennial and generation Z digital marketing. J Glob Scholars Market Sci. https://doi.org/10.1080/21639159.2020.1808812

Naumovska L (2017) Marketing communication strategies for generation Y—millennials. Bus Manag Strat. https://doi.org/10.5296/bms.v8i1.10260

Reinartz W (2018) The impact of digital transformation on the retailing value. Int J Res Mark. https://doi.org/10.1016/j.ijresmar.2018.12.002

Smith KT (2011) Digital marketing strategies that millennials find appealing, motivating, or just annoying. J Strateg Mark. https://doi.org/10.1080/0965254X.2011.581383

Soto-Acosta P (2020) COVID-19 pandemic: shifting digital. Inf Syst Manag. https://doi.org/10.1080/10580530.2020.1814461

Spilotro CE (2016) Connecting the dots: how IoT is going to revolutionize the digital. Digital USD. https://digital.sandiego.edu/honors_theses/25

Tan CN (2019) Determinants of green product buying decision among young consumers in Malaysia. Young Consum. https://doi.org/10.1108/YC-12-2018-0898

Chapter 11
A Study on Impact of Exchange Rate of Currencies on Indian Stock Market

H. Meena, Aishwarya Shiva Kumar, and Sandesh Bhat

Abstract The stock market plays a significant role in the economy, providing a platform for companies to raise capital and for investors to invest in profitable opportunities. The National Stock Exchange (NSE) is one of the leading stock exchanges and is considered as the backbone of the country's capital market. The NSE is an avenue for trading securities, including equities, debt and derivatives, and provides an efficient and transparent marketplace for investors and traders. This research study aims to examine the relationship between the Indian stock market (NSE) and a basket of currencies, including the US dollar, Euro, Japanese Yen, Great Britain Pound and Chinese Yuan. The study employs a time-series econometric analysis using weekly data over a period of fifteen years from January 2008 to February 2023. The study examines the impact of exchange rate fluctuations on the NSE. The research findings provide important insights for investors and policymakers in understanding the dynamics of the Indian stock market and its relationship with the global economy. The results of this study could have implications for investment decisions, risk management and policy formulation in the context of the Indian economy.

Keywords Stock market · NSE · Exchange rate · Investment decisions

11.1 Introduction

The Stock market is a place where the shares of a listed company are traded. It is a market for financial securities, i.e. those which have been already issued and listed on a stock exchange. These securities are purchased and sold continuously among the investors without the involvement of the companies. The Stock market is observed as an investment path in the modern era, but stock market investments are subjected

H. Meena · A. Shiva Kumar
Malleshwaram Ladies' Association—Academy of Higher Learning, Bengaluru, India

S. Bhat (✉)
Commerce and Management Department, MLA Academy of Higher Learning, Bengaluru, India
e-mail: sandeshrbht@gmail.com

A. Edward J. et al. (eds.), *Digital Transformation for Business Sustainability*, Contributions to Environmental Sciences & Innovative Business Technology,
https://doi.org/10.1007/978-981-99-7058-2_11

to market risk as it is volatile in nature. Because of fluctuations, it becomes difficult to safeguard the principle amount invested. People are fond of investing in the stock market due to the high returns, it is capable of yielding when compared with other investment avenues. To make proper investment decisions one should analyse various factors that could possibly have an impact on the stock market. We have considered the National Stock Exchange (NSE) in our study among the 23 recognised Stock exchanges of India. We have considered NSE as it is the most widely traded platform in the country and the other 21 are Regional Stock Exchanges. With proper decisions and investments one can get high returns in the Stock market and hence it is essential to understand the factors that influence the highs and lows of the stock market and affect the growth of our investments (Luo et al. 2006).

The NSE is the largest stock exchange in India by market capitalization, with over 1,600 companies listed on its platform. The exchange operates on a fully automated trading system, enabling investors with real-time access to market information and trading opportunities. The NSE has been at the forefront of several market reforms, including the introduction of electronic trading and settlement systems, which have helped to enhance market efficiency, transparency and investor protection.

The research aims at analysing the NIFTY returns to that of the Exchange rates of US dollar, Euro, Japanese Yen, Great Britain Pound and Chinese Yuan through which one can study the fluctuations of the NSE to make effective investment decisions based on the fluctuations on the currency market. The scope of the research extends towards establishing the impact of variables on NSE NIFTY. This research paper depicts the relationship between the stock exchange, which is the National Stock Exchange (NSE), to the currency exchange rates around the world (Mahapatra and Bhaduri 2019).

11.2 Review of Literature

The study conducted by Luo et al. (2006) has taken into consideration of Foreign currency impact on the value of the New Zealand public listed companies using New Zealand/US exchange rate and Trade weighted Index factor return. The model used in this Study is Augmented market model. The hypothesis of the study being (a) Foreign currency exposure is a function of firm's size and its industry affiliation. (b) Foreign currency exposure is a function of financial indicators, such as dividend yield, liquidities and P/E ratio. The study has resulted in very weak and ambiguous evidence of the exposure of foreign currency exposure on the value of New Zealand companies.

This research focuses on the dynamics of impact of the currency fluctuation on Indian Stock market during the period of 2005–2016 before the occurrence of Financial crisis (Mahapatra and Bhaduri 2019). The study has used a two-factor arbitrage pricing model which has presented that the Stock market had reacted significantly towards the fluctuation in the foreign exchange rates. The interpretation has been stated such that the Indian investors had increasingly expected a risk premium on the investments made adding to the exposure towards exchange rate risk.

11.3 Statement of Problem

Foreign exchange fluctuation has a great impact on the economy and financial status of a nation. There are various factors which affect the Nation's Finances where the Exchange rates are one of the factors which has a greater influence towards international trade and operations. Currently all the nations are interdependent on each other for resources, in this situation the countries trade with the help of exchange rate and basket of currencies. With fluctuation in these exchange rates comes the change in demand and supply of goods and services. Hence this study aims at studying the impact of currency fluctuation of the present Basket of currency towards the Indian Stock Market.

11.3.1 Objectives of Study

- To study the Investment pattern of investors against currency fluctuations in India.
- To study the impact of currency fluctuations on the Indian Stock market.

11.3.2 Significance of the Study

The study aims to establish the relationship between the currency rates of the Basket of currencies (US Dollar, Japan Yen, European Euro, Chinese Yuan, Great Britain Pound) to that of the variations occurring in the Indian Stock Market. The Indian stock market is a platform where the majority of investments are made by the citizens of India. The fluctuation in these markets affects the investments of the people hence it is essential that the investors are rational and take right decisions before investing in the Stock market. In order to make the right investment decisions, one has to have access to current affairs and factors that could possibly affect the investments made. With this study, the investors will be able to analyse the extent of influence that the exchange rates impact on the fluctuations in the Indian Stock market.

11.4 Research Methodology

Research methodology involves preparation of broad action plan or blueprint of this study; this is mainly house research and includes the following.

11.4.1 Sampling Procedure

For the study, there is no primary data which is used in the research. Secondary data has been collected from investing.com. The secondary data includes information related to Indian Stock market returns in particular to NSE and Exchange rates (USD, JPY, GBP, EURO, CNY) for a period of 15 years from January 2008 to February 2023.

11.4.2 Source of Data

- The data is collected through various sources.
- Secondary data has been collected to obtain additional information to support the study.
- The models used for analysis are ARIMA (Auto Regressive Integrated Moving Average) and ARCH (Auto Regressive Conditional Heteroskedasticity).

11.5 Data Analysis

See Tables 11.1 and 11.2.

Table 11.1 Analysis using ARIMA

ARIMA	P-value	Akaike criterion
USD/INR	2.88e−035***	10640.02
GBP/INR	0.7601	10777.62
EUR/INR	0.0101**	10771.13
JPY/INR	1.45e−025***	10690.13
CNY/INR	1.04e−018***	10705.13

Table 11.2 Analysis using ARCH

ARCH	P-value	Alpha(0)	Alpha(1)	Akaike criterion
USD/INR	3.39e−034***	3.69e−040***	6.15e−07***	10546.56
GBP/INR	0.2556	2.78e−040***	5.52e−07***	10695.03
EUR/INR	0.0002***	5.15e−044***	1.76e−07***	10681.89
JPY/INR	1.29e−032***	9.76e−040***	9.33e−09***	10571.68
CNY/INR	2.66e−020***	7.16e−040***	3.36e−07***	10611.48

*** - 99% LOS

Analysis: Both ARIMA and ARCH analysis tables indicate that the currencies USD, EUR, JPY and CNY have the P-value level less than 0.05. Whereas, the currency GBP has a P-value more than 0.05.

11.6 Findings

- H0 = The variables do not have a significant relationship with the Indian Stock market (NSE).
- H1 = The variables have a significant relationship with the Indian Stock market (NSE).
- The P-value level of the currencies USD, EUR, JPY and CNY are less than 0.05. Thus, we reject H0 and accept H1. This indicates that there is a significant relationship between the four variables and NSE.
- The currency GBP has a P-value of more than 0.05. Therefore, we accept H0 in case of GBP and reject H1. This indicates that there is no significant relationship between GBP and NSE.
- The investment risk in the NSE Stock market can be reduced by studying the exchange rate fluctuations which can possibly affect the Stock prices.
- The relationship between variables should be considered to make proper investment decisions while investing in the Stock Market.

11.7 Conclusions

In today's world, the number of investments in the Stock Market is rapidly increasing. Hence the returns observed in NSE NIFTY do affect the Indian economy. The Stock market has witnessed ups and downs through financial crises and Stock market crashes, due to various reasons. This gives an opportunity to analyse the impact of the currency fluctuation on Indian Stock market NSE NIFTY.

The analysis states that there is a significant relationship between USD, JPY, EUR and CNY on NSE, while GBP has no significant relationship with NSE. Hence any change in the pattern of USD, JPY, EUR and CNY has an impact on the NSE returns. The conclusion can be drawn as any changes in the pattern of exchange rates of USD INR, JPY INR, EUR INR and CNY INR have an impact on the NSE NIFTY, whereas any change in the pattern of GBP INR does not have any impact on the NSE returns. The investors can take into consideration these variables while investing in the Stock market and accordingly reduce the risk on investment The study considers the basket of currencies for its widely accepted range around the world. The study considers only currency impact on Stock market and does not consider any other macroeconomic variables. The study does not build a model to estimate Stock returns based on currency fluctuation. The investment risk in the NSE Stock market can be reduced by studying the exchange rate fluctuations which can possibly affect the Stock prices. The relationship between variables should be considered to make proper investment decisions while investing in the Stock Market.

References

Luo HR, Visaltanachoti N, Kesayan P (2006) Analysis of foreign currency exposure of the New Zealand stock market. Invest Manag Financ Innov 3(1):132–141

Mahapatra S, Bhaduri SN (2019) Dynamics of the impact of currency fluctuations on stock market in India: assessing the pricing of exchange rate risks. Borsa Istanbul Rev 19(1):15–23

Chapter 12
Blended Learning as a Sustainable Strategy in Digital Transformation—A Study of Challenges Faced by Parents at Sultanate of Oman

Alamelu Mangai, Jaffer Ali Khan, and Samiulla N. Mallik

Abstract This paper explores the concept of blended learning as a sustainable strategy in facilitating digital transformation within organizations. It begins by analyzing the driving forces behind the shift toward digitalization and the challenges that necessitate a balanced learning approach. The study delves into the core principles of blended learning, highlighting its adaptability and effectiveness in fostering employee development and organizational learning. The study explores the changing dynamics of parental involvement in the learning process due to blended learning. It examines the shifting responsibilities and roles parents are expected to play in facilitating their children's education, both inside and outside the traditional classroom.

Keywords Blended learning · Parental involvement · Changing dynamics

12.1 1 Introduction to Study

Due to the circumstances of Covid-19, Sultanate of Oman has decided to implement blended education in all the schools and colleges. Discussions, meetings, and communication with parents about teaching methods, the Supreme Committee in charge of looking for a mechanism to deal with Covid-19 chose blended education as the best solution for educational continuity. Furthermore, the Sultanate has tended to implement this type of education due to its interest in public health and limiting the spread of the virus. Some parents are dissatisfied with this educational method. They are unsure whether it will be implemented. Face-to-face learning is

A. Mangai (✉) · J. A. Khan
University of Technology and Applied Sciences—Al Mussanah, Al Mussanah, Sultanate of Oman
e-mail: alamelu@act.edu.om

S. N. Mallik
University of Technology and Applied Sciences—Shinas, Shinas, Sultanate of Oman

A. Edward J. et al. (eds.), *Digital Transformation for Business Sustainability*,
Contributions to Environmental Sciences & Innovative Business Technology,
https://doi.org/10.1007/978-981-99-7058-2_12

superior to blended learning because it involves a strong interaction between teacher and student and allows the teacher to see immediate feedback from students. The family's role in distance education differs greatly from its role in traditional education, including follow-up and evaluation (Vasquez 2020). The role of the family in the traditional education system is limited to monitoring the development and progress of the student's ability to achieve academic achievement based on the curricula, educational, and entertainment activities that the student participates in during the school day. Distance education imposes a pattern that differs from the traditional role of the family and places an increased burden on parents. The challenges and problems that parents face during blended learning are examined in this study, as is the new teaching style that parents must adopt. The research will assist parents in assessing the challenges and devising a solution that will allow them to provide better educational opportunities for their children. Parents have always been extremely important in their children's emotional and intellectual development. They are their children's first teachers, providing them with the footholds they need to adjust to school life. As the coronavirus pandemic halts face-to-face learning, parents find themselves in a difficult position. Parents now have the important task of ensuring that their children receive a high-quality education while remaining safe. Although education suffers as classrooms are forced to close their doors to eager students, countless parents are stepping up to support their children as they adjust to the new learning environment (DepEd 2020).

These developments have sparked several questions, including what kind of instruction will be provided at home, how parents will oversee their children's education, especially with young children, and what class schedule will be adhered to by parents who work full-time. Many issues have been raised, including how parents will support their children's education while still seeking ways to make money. These parents are willing to adapt to the new routine but have legitimate worries. Nonetheless, teachers will continue to support children's learning by acting as ever-ready learning advisors; therefore, parents need not be concerned. They will also help parents and children complete online and offline modules and workbooks. If they still need to complete tertiary education, how will they be able to teach their children? As a result, blended learning is now a successful method for teaching what children need to know, and parents and guardians are crucial in ensuring that their children continue learning despite the anticipated setbacks. Some parents consider this an opportunity to monitor their children more closely and become more active in their education. This illustrates that parents must be more apprehensive about ensuring their children's education remains a primary priority despite the Covid-19 problem. We grew up and developed together as a group. Education should be a joint effort between teachers and parents to maximize a learner's overall growth and development. Their cooperation can encourage students to push past their comfort zones and feel secure during their educational journey. To address these issues and enhance parental engagement in blended learning, the study proposes actionable recommendations and strategies. These include targeted training programs for parents to improve their digital literacy, measures to bridge the digital divide, and initiatives to foster a more inclusive and supportive learning environment. In conclusion, this study provides valuable insights into the challenges parents confront in the context of blended learning and digital transformation in education (Li and Lalani 2020).

12.2 Statement of the Problem

Blended learning for children in grades one through four has placed parents in a new position that requires them to take more responsibility for their children's education. The family must accept the new approach to education in Oman and adjust to the new situation. However, the new approach places parents in a position where they must deal with any problems that arise during blended learning. Many of them oppose the new approach because of the numerous challenges that will arise during implementation.

12.3 Aims and Objectives of the Study

- Identify the obstacles that parents face when implementing blended education for their children.
- How do parents view blended learning strategies during a pandemic?
- What effects do parents perceive the blended learning strategy to have on their children's learning?

12.4 Significance of the Study

This is an important study that has not been conducted previously by Omani researchers. Many studies have focused on students and teachers, so this study is unique in that it only looks at parents. It will benefit both parents and students. The study will assist parents in identifying and highlighting challenges. It has aided families in adapting to a blended learning approach, which may aid students in achieving higher academic levels.

12.5 Review of Literature

Interactions between students and teachers are crucial because they are linked to students' inability to complete online courses. According to Duterte (2020) study, which was quoted in this paragraph, peer and family support for learners is crucial for success in both online and face-to-face learning (Internet Society 2017). Learners from various backgrounds demand assistance in web-based courses that may come from family, friends, co-workers, or classmates. Peer support helped new computer users and programs.

By offering more options for cutting-edge learning experiences, blended learning as a pedagogical innovation seems to extend students' horizons in the classroom. Also, it enhances their academic performance (Mina et al. 2020). As a result, it can be utilized at many levels of education as a kind of online or e-learning. However, university students can use it more effectively because they are adults who are self and socially motivated by their own choice and pace of study (Llego 2020). In a 2009 study, Melton, Bland, and Chopak-foss evaluated students' learning outcomes and satisfaction with blended course delivery to conventional classroom instruction in health education. The results revealed a significant difference between classroom (49.788) and mixed (54.986) satisfaction ratings ($p = 0.01$). Students who received mixed education appeared to have much better overall scores than their counterparts ($p = 0.048$). According to the survey, students favored blended course delivery over conventional classroom lectures. Similarly, Davidson (2009) reported a change in instructional design that involved team learning and blended learning in teaching an undergraduate medical course. At the University of Central Florida, Duterte (2020) examined how organizations, academics, and students regarded blended learning. They sought to ascertain the effects of blended learning on overall study satisfaction. The findings showed a significant difference ($p < 0.01$) between the satisfaction levels for the classroom (49.788) and mixed (54.986) instruction. Students who received blended teaching appeared to have considerably better overall grades than their peers ($p = 0.048$). According to the survey, students preferred mixed course delivery to traditional classroom lectures (Duterte 2020).

Similarly, Davidson (2009) reported a revolution in teaching an undergraduate medical course that included blended and team learning, which emerged as an innovation in instructional design. However, Black (2009), at the University of Central Florida, examined how organizations, academics, and students perceived the adoption of blended learning. Overall satisfaction with the study was found by Black (2009), who set out to determine the impact of blended learning on fifth grade students' achievement and development of science operations.

The most important results of the study showed the effectiveness of blended learning because it combines e-learning and traditional learning. Further, it assisted in providing educational materialism in many ways through a statistically significant difference between the marks of the experimental group who studied via blended learning. The degrees of the control group who studied via traditional learning in the post-te (Black 2009) sought to determine how blended learning affected sixth graders' immediate and delayed achievement in their English language course. A quasi-experimental design was used to accomplish the study's goals, and the researcher created an achievement test whose validity and reliability were evaluated. This study involved (60) pupils from Quwaismeh's Um-Qasir School for Boys. A study on blended learning was conducted by Black (2009). It examined how employing blended learning affected students' achievement and information retention for biology students in the fifth grade. Two parts of a high school in the heart of Baghdad were randomly selected, one representing the experimental group that studied using blended learning and the other representing the control group that was studied using traditional learning. The overall sample size was 60 students after statistically deleting five failed pupils, with each group consisting of (30 students). A 30-item multiple-choice achievement test was created, and the study's validity and reliability were examined. The outcomes demonstrated that the experimental group outperformed the control group in achievement and information retention tests.

Face-to-face experiences are included in blended learning setups, and learner's positive attitudes toward such sessions may indicate blended learning effectiveness. The learners prefer face-to-face instruction because it facilitates social interaction and communication skills learned in the classroom. Their preference for the online session was limited to the extent that it supplemented traditional face-to-face learning. In this study, learners had positive perceptions of blended learning but preferred face-to-face instruction with step-by-step instructions.

Some learners prefer face-to-face engagement in the blend because, according to Black (2009), they are more effective when they connect with teachers and classmates in person. The authors also compared online and on-campus learning, whereas our study integrates both, stressing the face-to-face component. Research contrasting blended learning with conventional face-to-face instruction has indicated that learners perform equally well in both settings, regardless of the delivery strategy. Another study discovered that learning experience and performance were enhanced when traditional course delivery was paired with online learning. Interactive one-to-one discussions, quick warm-up questions, summary exercises, and evaluations were the three assessment approaches covered by Black (2009). Different questions offer individualized feedback while reiterating important information during the learning process.

12.6 Methodology

12.6.1 Study Design

A descriptive research approach was used in this study. Descriptive research is a fact-finding investigation that entails adequate and accurate interpretation of data. The specific current condition was described using descriptive research. In light of the study's goal to describe the difficulties parents encounter with blended learning, the methodology is generally adequate. The normative survey strategy and evaluation, frequently used to investigate viewpoints according to respondents who may accurately represent the entire population, were employed in the descriptive research. The survey is suited for this study since it allows for the formulation of generalizations by the researcher.

12.6.2 Sample Design

A convenient sampling technique was adopted to collect samples from the population.

12.6.3 Study Sample

The study sample consisted of 120 parents who were randomly chosen for the study.

12.6.4 Tools for Data Collection

The study depends on the primary data, special care was taken to design the questionnaire. The questionnaire method was used to collect primary data from the parents. Secondary sources of data were collected from books, journals, and websites for identifying the population and sample.

12.6.5 Reliability Analysis

Reliability is brought to the fore when variables derived from summated scales are utilized as predictor elements in objective models. It is crucial to understand whether the same set of items would elicit the same replies if the same respondents were asked the same questions repeatedly, as summated scales are an assembly of connected items intended to measure underlying constructs. They claim to be dependable only when the test variables produce consistent and reliable results during multiple test administrations. Confirmatory Factor Analysis was used to examine the validity of the groups in the survey. A reliability of 0.8 (Cronbach alpha) implies that the variables used for the study and the grouping of variables are reliable. It also shows that the instrument's reliability is suitable for data collection. The following construct reliability table shows the findings of this study (Table 12.1).

Table 12.1 Reliability and internal consistency

Construct	Reliability (Cronbach's alpha)	Internal consistency
Blended learning—challenges faced by parents	0.8	Better

12.7 Results and Discussion

Between the ages of 25 and 30 ($f = 25$; $\% = 21$); 31–36 ($f = 35$; $\% = 29$); 36–40 ($f = 30$; $\% = 25$); 41–45 ($f = 25$; $\% = 21$); 46–50 ($f = 5$; $\% = 4$). All age groups were well represented among respondents as they came from all age groups. There are 38 respondents on average, with an average age of 38. Parental involvement in this might indicate that they are under the age of "prime working age" (Table 12.2).

In the table showing respondents' educational backgrounds, 25% of respondents are males with a bachelor's degree and 33% of females with a diploma (Table 12.3).

Responses Regarding the Blended Learning Approach and Their Attitudes and Perceptions

The parents' attitudes and perceptions toward the blended learning approach. Based on the mean value of 3.95, we can conclude that respondents recognized the value of blended learning for all of the students. It gives more benefits to the children' is 3.93 and the blended learning is useful for all mean value is 3.84, the mean value is neutral in terms of 2.95—Easy to adopt, adoptable 3.63, economical is 3.15 and useful to learn is 3.10. However, respondents have neither positive nor negative perceptions of the blended learning approach's other identified characteristics. However, this would imply that parents have a favorable attitude toward blended learning. Blended learning is not opposed by parents, but they are not enthusiastic or well informed about the format (Table 12.4).

Table 12.2 Parents' demographic profiles

Age	Frequency	Percentage
25–30	25	21
31–36	35	29
36–40	30	25
41–45	25	21
46–50	5	4
Total	120	100

Table 12.3 Educational details

Education	Male	%	Female	%
Diploma	25	42	20	33
Undergraduate	15	25	10	17
Postgraduate	10	17	5	8
PhD	5	8	5	8
School level	5	8	20	33
Total	60	100	60	100

Table 12.4 Blended learning approach and the perceptions of parents

Statement—blended learning	Mean	Rating
Easy to adopt	2.95	Neutral
More benefit to the children's	3.93	Positive perception
More effective	3.84	Positive perception
Useful for all	3.95	Positive perception
Adoptable	3.63	Neutral
Economical	3.15	Neutral
Easy to learn	3.10	Neutral

Table 12.5 Frequency of blended learning approach and the perceptions of parents

Statement-blended learning approach	Frequency-yes	%	Frequency-no	%
Enrich the knowledge	25	21	20	17
Motivate students to learn more	13	11	35	29
Adopting technology easily	30	25	35	29
Effects on the grades of children	28	23	38	32
Increase more responsibilities	25	21	12	10
Virtually connected outside	32	27	23	19
Facing challenges for the children	12	10	10	8

The table implies, that respondents perceived blended learning. 27 respondents agreed that they are virtually connected outside, 25% respondents are adopting technology easily and affects the grades of the children, 32% respondents, mentioned that blended learning affects their grades of the children, and 29% adopt technology easily and motivates students to learn more. The results showed that most parents believe the blended learning approach has a positive impact on their children's learning, but they also believe it has difficulties for them. The parents' satisfaction and optimism about the learning approach could be attributed to it (Table 12.5).

12.8 Conclusion

The study involved computing and analyzing the collected data for this study, titled "Blended Learning—A Digital Transformation Study of Challenges Faced by Parents at Sultanate of Oman" descriptive statistics were used. Our study was designed to identify the parents' age, highest level of education, income, and occupation based on their demographic profiles. In addition, the study determined whether blended learning had an impact on the children's learning and how their parents perceived it. According to the weighted mean of 3.5, the respondents' attitudes and perceptions of blended learning were neither positive nor negative. According to this study, blended learning is viewed positively by parents. The majority of respondents are optimistic about their children's learning outcomes as a result of the blended learning approach. A positive response was received from parents regarding the blended learning approach. Blended learning is perceived and regarded as neither positive nor negative by parents and teachers. Some of their recommendations favor blended learning. Many parents are satisfied with the impact of blended learning on their children's education. Pedagogical approaches have resulted in positive feedback from the parents. The school must formulate a plan to assist the parents in assisting their children's learning at home since most of the parents were diploma and undergraduate holders. To ensure students receive a quality education, schools must properly train parents on how to implement the blended learning approach. Parent perceptions of the blended learning approach are very positive, so the school aims to foster positive engagement from both parents and students through the best school services and activities.

References

Black EW (2009) An evaluation of familial involvements' influence on student achievement in K-12 virtual schooling. Unpublished dissertation, University of Florida, Gainesville, FL

Davidson L (2009) Educational innovation in an undergraduate medical course: implementation of a blended e-learning, team-based learning model. 1–161. https://qspace.library.queensu.ca/bitstream/1974/1667/3/Davidson_L_K_200901_MEd.pdf

DepEd (2020) Official statement department of education. https://www.deped.gov.ph/2020/09/21/deped-reinforcescrisis-resilient-strategy-through-the-cvif-dynamic-learningprogram

Duterte R (2020) No vaccine, no school opening. Philippine Daily Inquirer. https://newsinfo.inquirer.net/1281442/duterte-no-vaccine-noschool-opening. Abelardo L, Lomboy M, Lopez C, Balaria F, Subia G (2019) Challenges encountered by the national high school teachers in doing action research. Int J Engl, Lit Soc Sci (IJELS) 4(4). https://doi.org/10.22161/ijels.4418. ISSN 2456-7620

Internet Society (2017) Internet access and education: key considerations for policymakers. https://www.internetsociety.org/resources/doc/2017/internet-access-and-education/#_edn1

Li C, Lalani F (2020) The COVID-19 pandemic has changed education forever. https://www.weforum.org/agenda/2020/04/coronavirus-education-global-covid19-online-digital-learning

Llego MA (2020) https://www.teacherph.com/deped-online-learningsmartphone-education/

Mina J, Subia G, Pascuala P, Tuliao R, Pastorfide D (2020) Inclinations of engineering and marketing management students to engage in online learning technology amidst the COVID-19 pandemic. Technol Rep Kansai Univ 62(09). ISSN 04532198

Vasquez D (2020) https://businessmirror.com.ph/2020/04/18/online-educationduring-a-pandemic

Chapter 13
A Study on Work Life Balance of Teaching Women Employees Through Training Programs in Government First Grade Colleges in Chitradurga District

Syeed Rajeena and H. S. Jamuna Rani

Abstract The study employed a mixed-method approach, combining quantitative surveys and qualitative interviews to gain comprehensive insights into the experiences of teaching women employees. A structured questionnaire was administered to a diverse sample of teaching women employees across various Government First Grade Colleges to gather quantitative data. Additionally, in-depth interviews were conducted with a selected group of participants to delve into their perspectives and experiences in greater detail. The research focused on several key aspects, including the challenges faced by teaching women employees in maintaining work-life balance, the impact of work-life balance on job satisfaction and performance, and the perceptions regarding the effectiveness of existing training programs aimed at enhancing work-life balance. The work-life balance of teaching women employees are more important concept because, after covid 19 in the education system, huge changes are going on in offline teaching converted into an online system. In this situation, the negative vibration starts from the side of teaching women employees, learning E-tools to reach the students and without training, it is very difficult to operate E-tools, so they expect provide to E-tools training from the side of institutions. To continue in the teaching field and sustainable development in their profession, this study identifies the work-life balance of teaching women employees in government first-grade colleges in the Chitradurga district and how they managed themselves in these new challenges. And giving guidance to them on how they can balance their work with sustainable development in the education field.

Keywords Work-life balance · Training programs · Government first-grade colleges · Educational institutions · Teaching women employees · Chitradurga district

S. Rajeena · H. S. Jamuna Rani (✉)
CMR University, Bangalore, India
e-mail: phdjamunarani@gmail.com

A. Edward J. et al. (eds.), *Digital Transformation for Business Sustainability*, Contributions to Environmental Sciences & Innovative Business Technology,
https://doi.org/10.1007/978-981-99-7058-2_13

13.1 Introduction

Teachers and society are interrelated concepts and education play a vital role in society. At the same time, training teaching women employees in the field of education is needed because students' development is the ultimate goal of a future-oriented society, the gradation of society and education in society and constant economic growth so that the teacher is pillars of the society they guide students to follow the ethics in the society and do the work like responsible citizens, if the teacher learns how to handle the situation they sow the seeds of knowledge in the mind of students then its roots will be deeply penetrated in the society (Arunima and Nangia 2020).

From that point of view, we think about teachers, but it is very difficult for women employees. They have to balance both personal and professional. So, if they get proper facilities from the side of institutions, they will continue to do their job, otherwise, they think about not being interested in doing work, The sustainability of teaching women employees is very important. Education is a significant factor in the progress of a nation. Human being life starts with proper education. Without education, the development of a nation is very difficult, and providing quality teaching and teaching facilities is a very common responsibility of government institutions and organizations. The institutions may think that the teaching role is very important, The teacher always transfers all the qualities to future generations, A well-trained teacher can offer this responsibility easily, so teaching women employees to provide training is very important (Jadon and Shrivastava 2018).

Building the nation is not only in the hands of men, it requires female contribution because in the world 50% of women in there, but overall development purpose participation of women is also needed (Shankar and Jose 2016). In HR we have considered people are assets of society because their ability, innovation ideas, creativity in their work, it increases productivity in the organization. Today, education is the front line of worldwide discussion and decisions to overcome the development of a nation's economy, politics, and society through the education system, to face the challenging world to create citizens of tomorrow is an unsettling task for teachers (Bali et al. 2021). The educational institution depends upon the central and local government policy, to make the curriculum for teaching, methodology, students, parents, and institutions. These are major tasks for teachers in their work and life. The teacher plays a key role in shaping the world for tomorrow (Wolor 2020). The concept of training and development in teaching women employees is a burning issue in the educational sector. Due to globalization many employees work in different shifts, which results in having less time spent with family and friends (Depthy and Toppa 2020). During the situation of Covid-19, the teaching faculties have teachers and society are interrelated concepts and education plays a vital role in society. At the same time, training teaching women employees in the field of education is needed because students' development is the ultimate goal of a future-oriented society, the gradation of society and education in society and constant economic growth so that the teacher is pillars of the society, they guide students to follow the ethics in the society and do the work like responsible citizens, if the teacher learns how to handle the situation they sow the seeds of knowledge in the mind of students then its roots will be deeply penetrated in the society (Hashimi 2021).

From that point of view, we think about teachers, but it is very difficult for women employees. They have to balance both personal and professional. So, if they get proper facilities from the side of institutions, they will continue to do their job, otherwise, they think about not being interested in doing work, The sustainability of teaching women employees is very important. Education is a significant factor in the progress of a nation. Human being life starts with proper education. Without education, the development of a nation is very difficult, and providing quality teaching and teaching facilities is a very common responsibility of government institutions and organizations. The institutions may think that the teaching role is very important, The teacher always transfers all the qualities to future generations, A well-trained teacher can offer this responsibility easily, so teaching women employees to provide training is very important (Kosanovic 2021).

Building the nation is not only in the hands of men, it requires female contribution because in the world 50% of women in there, but overall development purpose participation of women is also needed (Yadav and Jandan 2021). Today, education is the front line of worldwide discussion and decisions to overcome the development of a nation's economy, politics, and society through the education system, to face the challenging world to create citizens of tomorrow is an unsettling task for teachers. The educational institution depends upon the central and local government policy, to make the curriculum for teaching, methodology, students, parents, and institutions, These are major tasks for teachers in their work and life.

The teacher plays a key role in shaping the world for tomorrow. The concept of training and development in teaching women employees is a burning issue in the educational sector. Due to globalization, many employees work in different shifts, which results in having less time spent with family and friends (Jandan 2021). During the time of COVID-19, the teaching faculties have moved from offline teaching to online teaching, and they are facing new challenges in learning ICT and decreasing academic activities, to the development of educational institutions in globalization adopting policies like basement standards and curriculum instruction. They have to adopt government policies toward teaching women employees to increase the productivity of the institutions to motivate the employees. The concept of work-life balance of female employees has multiple problems that are faced by the time a husband earns and the wife cooks and stays at home, but now the time has changed, the woman also earns but she does not get any excuse for the home chore. To create their own identity and make a name in society and also create plot forms in organizations, Indian women are struggling more compared to their male counterparts. The present scenario, educational institutions provide training, and career development, that will create more opportunities for women to enter into the field of job. The work-life balance of teaching women employees is part of an educational institution and is linked with the life and work of the employees, and it's not similar to all, it is different from one to another.

In the present scenario, the issue of work-life balance is more important. In India, it has a male-dominated society, exclusively males are involved in all progressive activities and not domestic work, Child bearing, and child-rearing are the sole occupations of women, Women in India have not enjoyed a good status historically, they have more burdened with work of all sorts in their life if they do work outside they suffer from discrimination and exploitation in this male-dominated society. After changing economic conditions and social demands that changed work throughout the world, the concept of a work-life balance was relevant in a working environment.

13.2 Review of Literature

Hashimi (2021) stated in this study that the effect of technology during Covid-19 became very crucial in almost every sector globally. Due to this crisis transformation, educational institutions have caused huge challenges, which push the teachers to skill themselves toward teaching methodology to bring innovations. For educational programs to be efficient, they need to refresh their academic affairs, so that computerized complement the teachers to maintain the quality of education.

Wolor (2020) discussed any one can enjoyable and stress-free life balance is essential for the work-life balance is important in every female teacher. They must have a good work-life balance because they effectively teach for their job satisfaction and to motivate the students.

Depthy and Toppa (2020) discussed in this study women working from any class are empowered themselves and they have reached a new platform, and they have created their self-identity, freedom, personal and professional. And now, they get in a better position and achieve their career objectives, In this dual-career, women employees are suffering to balance their work and life. Comparison of male and female workers are involved more in domestic duties, the majority of female workers are dissatisfied with organization policies and work schedules, and married working women face a lot of problems in their personal and professional lives. Only some female workers have traits that are successful in their work and life. In this study, they are given suggestions for implementing some human resources policies for the benefit of working women.

13.3 Statement of the Problem

The main focus of the paper is the work-life balance of teaching women in government first-grade colleges, who are in an alone-ring battle position to keep a balance between personal and professional life. The dream of being independent and career orientation became a tough challenge for the teaching woman. She had to perform the duties of a perfect homemaker in the family and a courageous teacher in college. Whether someone lends help or not, being a mother, it is the woman's primary duty to take care of her children. At the same time, she has to manage her parents and in-laws, which is also difficult for Indian married women. Therefore, it is necessary to examine the differences between problems and training. It assists the organization in giving better sustainability for teaching women employees to attain a higher degree of job performance and enhance their satisfaction.

The dynamic in the job, continuous learning, long working hours, lack of consistency in work schedules, pressure and development of both non-technical and technical skills, professional competencies, family and financial obligations, and child and parental care have made a strong effect on balancing life and work of teaching women employees who are working in government first-grade colleges. Support from family is very crucial for the development of teaching women employees about family and work, otherwise, it creates conflicts in the family and disintegrates the family and social structure.

13.4 Need to Study

The concept of work-life balance is more important for teaching women. At the same time, identify the different factors which will affect their working situation, like as working hours, work-family conflict, psychological dis stress, work-family satisfaction, family responsibilities, child caring, office duties, career building, job satisfaction, also focus on how married women maintain their life and family responsibilities and if they will become mothers how they will manage their professional work, husband duties and caring of children. Women were beheld as caregivers or as homekeepers and were denied access outside the home. Today's women have made their mark in every field and are ready to take up challenges. As time changes, both husband and wife earn, but the wife still cooks, washes, and runs the house. So how does she balance herself in the field of teaching?

13.5 Objectives of the Study

- To study the teaching women employees work-life balance in government first-grade colleges in Chitradurga district.
- To understand the different training programs affecting the work-life balance of teaching women employees in GFGC.
- To examination of the different programs that impact the work-life balance of teaching women employees OF GFGC.
- To suggest various policies adopted for work-life balance of teaching women in GFGC Chitradurga district.

13.6 Scope of the Study

The study is limited to work-life balance of teaching women employees in government first-grade colleges in Chitradurga district.

13.7 Work-Life Balance of Teaching Women Employees in India

Women are the human resources of the nation, in that teaching professionals are the facilitators of information and ability through keen learning tools in the education field. In the current scenario, the speedy changes in the education system, expert teachers have adopted teaching practices in different styles through technical bases like online classes conducting YouTube videos uploading, and PPT presentations. Along with these, they were worried about their day-to-day work, stress, long time working mental stress, provoking students, controlling work, placement of students, conducting tests, doing assessment and counting workforce (Balamurugan 2020), reproach in India the females were generally used in non-managerial subordinate or low profile positions. But, day by day, their obligations were changed, and they moved toward acquiring all the positions in the current scenario. The contentions between fighting work demands and individual and family needs give off an impression of being the most conceivable explanation behind this situation of their work-life balance and their career development clashes.

13.8 Women in Teaching Profession

The success of educational institutions depends on the quality of teaching by the teachers and their ability. Teaching women's development is a censorious aspect of increasing teacher efficacy and satisfaction in the conditions of student learning. The good quality of teaching and the results for students depend upon the proper training of the teacher (Vandana Punia 2013). The education sector is the ultimate intellectual workplace. The development and exchange of ideas are constantly prepared in people's minds. Continuously giving lectures and attending meetings, doing academic work 9 to 5 busy schedules needed for achieving sustainable development. Balancing family and academics, especially for female employees, is a tough challenge, Women face dual roles and problems, academic responsibility, doing research, and career development, and fulfilling family demands them feel more stress, They need some strategies to balance both sides, family support, institutions support, to avoid barriers occur in their work and life (Aditi Aeron et al. 2015).

13.9 Training Programs for Teaching Women Employees

The work-life balance of teaching women employees in the education field shows the battle between work and life of their roles, it is an equilibrium between personal life and professional life. Maintaining balance is very important because it retains health and increases yield and satisfaction at work. In this competitive corporate world, many teaching women have an unbalanced life. Balancing both sides is a challenging task and a greater risk for teaching women facing ill health, and absenteeism and those employees with a better-balanced life live happier, more satisfied, and more productive lives at work and in their families. The training program on work-life balance for teaching women employees to practice instruments and approaches to developing a scattered balance between work and regular life.

The training provides them with the structure of tailor-made designed to individual requirements. Through training programs like group discussions, team games, activities, and role-plays based on learning, women employees gain the idea of various skills. The E-tools training, conducting Google Meet, Zoom Meet, how to upload files, create channels, PPT presentations, publications, conducting webinars, FDP programs, workshops, etc. So, these are training programs that motivate the teaching women employees in their teaching field to upgrade their knowledge and show their academic excellence.

13.10 Research Aim and Research Questions

The main aim of this research is to compare the work-life balance experiences of teaching women in the GFGC Chitradurga district. To balance their life and career properly, they need training programs from the side of institutions.

13.11 Research Question

- Employees have more pressure to work in the institutions. (a) Agree (b) strongly agree (c) Disagree (d) strongly disagree
- Employees have more pressure to work in the institutions. (a) Agree (b) strongly agree (c) Disagree (d) strongly disagree
- Employees need training: (a) Yes (b) No

13.12 Materials and Methods

A structured questionnaire will be used to collect the primary data.

Open access journals, research articles of previous studies, books on work-life balance, stress management, previous PhD thesis, etc. will be used to collect secondary data.

13.13 Data Analysis

- Descriptive statistics are used to know the demographic variable.
- Inferential statistics are also used to know the impact of variables.
- Microsoft Excel and SPSS will be used for the current study.
- Analysis of the work-life balance of teaching women employees of GFGC.

13.14 Interpretation

The findings of the Socio–Economic variables status are shown in Tables 13.1, 13.2, and 13.3, summarized as below.

1. It is inferred that the majority of the respondents are married. They are 70%.
2. Most of the respondents belong to the age group of 31–35 years.
3. Department wise, the commerce faculties are more. They are 48%.

Table 13.4 analyzed the respondents based on their educational qualifications. Majority of the teaching female employees qualify for Ph.D. 17 respondents are there.

Table 13.1 Socioeconomic status wise classification

Sl. no	Variable classification	No. of respondents	Percentage
1	Married	35	70
2	Unmarried	15	30
Total		50	100

Source Primary Data

Table 13.2 Age wise classification of the teachers

Sl. no	Age wise	No. of respondents	Percentage
1	25–30	11	22
2	31–35	18	36
3	36–40	09	18
4	Above 40	12	24
	Total	50	

Source Primary Data

Table 13.3 Nature of department wise classification

Sl. no	Departments	No. of respondents	Percentage
1	Arts	12	24
2	Science	10	20
3	Commerce	24	48
4	Management	14	28
Total		50	100

Source Primary Data

Table 13.4 Educational qualification of respondents

Educational qualification	No. of responses	Percentage
Post Doc	3	6
Ph.D	17	34
PG with M. Phil	20	40
PG with NET/SET	8	16
PG	2	4
Total	50	

Source Primary Data

13.15 Types of Training Programs

- Faculty development program
- Refresher training course
- Orientation program
- Webinar program
- ICT Tools training
- Workshop conducting
- Data analysis program
- SPSS Tools program
- Swayam course
- Other programs

Those who get training properly from this type of program have the ability to balance their work and life. Training will provide them with the confidence to handle the situation smartly without getting stressed, so these types of training programs motivate them to develop in their academic field.

Findings of the Study

1. Most of the respondents given opinion that their personal life suffers because of work stress.
2. Some respondents are happy with adopting work-life balance policies.
3. Many respondents say that institutions conducting faculty development programs are needed.

4. Employees feel that without a work-life balance, they have negative effects on both sides.
5. Getting training from the side of institutions teaching women employees to manage their work-life balance properly compared to not getting the training of women employees.

Benefits of Training Program

- Realize the Benefits of a work-life balance and identify the signs of an unbalanced life and the effects of it.
- Learning the techniques for a balanced lifestyle and communication effectively and drawing boundaries.
- Learning time management skills by setting the right priorities to develop their goal-setting skills.
- Balance and handle work and personal stress differently become more productive at their workplace by developing effective and smart work methods and improving life and health by creating a good balance.

The training program on work-life balance teaches women employees gets practice tools and techniques to develop a well-distributed balance in work and regular life. The training provides them with the structure of tailor-made designed on individual requirements. Through training programs like group discussions, team games, activities, and role-plays based on learning, women employees gain the idea of various types of skills and acquire knowledge.

Organizational Benefits

- Increased productivity and a decrease in absenteeism are noticed.
- The employees are happier and feel less stressed improvement in their health and well-being.
- Increased employees' loyalty, commitment, motivation better teamwork, and good communication between employees.
- Improved overall satisfaction and less emotional weakness at the workplace.

Personal-Benefits

- To get the training program the personal benefits are as follows.
- They reduce their stress by having greater support from the side of family members.
- Individuals get the time to pursue further education.
- The training program enables them to be involved in society and career development.
- They got better time management skills among themselves.
- Through training, they have an individual in their personal and professional life.
- The training provides confidence to maintain their health and welfare.

13.16 Conclusion

This study is to be considered as a psychological effect for teaching women of Indian culture. This study found that most of the women were working with a sense of missing life and institutions have not taken any steps to overcome them. In this situation, organizations should handle the work-life balance, balance among staff and provide basic training facilities to teach properly using ICT tools, adopt new policies according to this COVID situation and implement approaches to help their staff to find and manipulate their work-life balance. HRM department should help educational institutions to the establishment of a new performance evaluation strategy. Taking all these suggestions and recommendations will benefit educational institutions and teach women further to study the future scope Focus on understanding the preferences and behavior of different generations (e.g., Gen Z, Millennials, Gen X, Baby Boomers) in relation to sustainable products. This can help businesses tailor their strategies to target specific age groups effectively. Investigate the role of influencers and social media in shaping sustainable consumer behavior. Analyze how influencer endorsements and content impact consumers' choices when it comes to purchasing durable and eco-friendly products.

References

Arunima, Nangia R (2020) Work life balance—an integrative review of literature. JETIR 7(7)
Bali P, Sajnani M, Gupta RK (2021) Study on the work-life balance of hotel employees in metro cities. 12(13):3720–3728
Depthy, Toppa C (2020) College teachers' work-life balance. Int J Recent Sci Res 11(2):37647–37650. https://doi.org/10.24327/ijrsr.202011025150
Hashimi (2021) Female employees' work-life balance and impacting factors. Kardon J Econ Manag Sci 4(3):84–99. https://doi.org/10.31841/kjems.2021.101
Jadon A, Shrivastava S (2018) Women education in India: an analysis. 8(13). ISSN 2224-5766 (Paper). ISSN 2225-0484 (Online)
Kosanovic (2021) During a pandemic situation in covid 19 the work-life balance of teaching staff at higher education. Serban J Eng Manag 6(1):63–71. 10.59371sjem1010641
Shankar A, Jose S (2016) Work-life balance among women employees: a case study with reference to IT companies. Adel J Soc Work 3(1):16–29
Wolor CW (2020) The importance of work-life balance on employee performance millennial generation in Indonesia. J Crit Rev 7(9). ISSN 2394-5125
Yadav MJ, Jandan S (2021) Study of work life balance of women employees during work from home in the periodic covid-19. Int J Futur Commun Netw

Chapter 14
Way Forward for Sustainable Digital Marketing: A Bibliometric Analysis

Aparna Merin Mathew and M. P. Akhil

Abstract Digital marketing is the trending type of marketing that heavily uses the internet, Information, and Communication Technology (ICT) and online-based digital technologies to promote products and services. Internet made life much easier; a large number of customers spend their time on digital and social media for different reasons ranging from information gathering to final product purchase. Thus, marketers are allocating a sizable portion of their advertising budget to digital marketing. As a result, this paper sheds light on articles on digital marketing to identify top themes, determine the current state of research and studies in digital marketing, and indicate how influential works have shaped it. Using bibliometrics analysis, this study examined 987 papers published in Web of Science (WoS) database between 1993 and 2022. According to these findings, an average of 2.26 authors contributed to each paper on digital marketing, with a collaboration index of 2.49. USA, China, and UK are the top countries which are adding knowledge to the existing arena of digital marketing. The results depict a positive trend in academic and research interest in digital marketing from 1993 to 2022 as the annual scientific productivity shows a continuous increase over these years. Digital transformation and digitalization also act as an accelerator of the same.

Keywords Digital marketing · Bibliometric analysis · Digital transformation · Digitalization · Marketing

A. M. Mathew (✉)
All Saints' College, Thiruvananthapuram, Kerala, India
e-mail: lncs@springer.com

M. P. Akhil
Alliance University, Bengaluru, India

A. Edward J. et al. (eds.), *Digital Transformation for Business Sustainability*,
Contributions to Environmental Sciences & Innovative Business Technology,
https://doi.org/10.1007/978-981-99-7058-2_14

14.1 Introduction

The term Digital Marketing was introduced in the 1990s. Digital marketing, also referred to as online marketing, refers to all marketing activities conducted over the internet (Yasmin et al. 2015). Search engines, social media, email, and other websites are used by businesses to connect with current and prospective customers (Desai 2019). Text messages and multimedia messages are often used to communicate. Digital marketing is very cost effective when compared to other modes of marketing and has a tremendous commercial impact on the growth prospects of a business concern (Parsons et al. 1998). With the wide popularity and application of internet and ICT, it is not only just selling products alone, but also provide information about products, advertising space, software programs, auctions, stock trading and matchmaking (Chaffey and Ellis-Chadwick 2019). Digital marketing provides companies with databases that allow them to track customer information more effectively and develop and market products accordingly (Bala and Verma 2018, Morris 2009, Parsons et al. 1998). Given that digital marketing is a major segment and an expanding area of marketing, there has not been a lot of scientometric inquiry into the topic (Aksnes et al. 2019; Moral-Muñoz et al. 2020). The most recent advancements in the field must be identified and tracked. This can be performed by using various pertinent and extensive keywords, together with the most reliable global indexing and extrapolating repository, Web of Science (WOS), and by studying hitherto undiscovered regions via bibliometric studies (Kraus et al. 2022; Aksnes et al. 2019; Murillo-Ramos et al. 2022; Akhil 2022; Rose et al. 2020). This method can be used to obtain significant and complete insights from digital marketing literature.

To cover a lacuna in digital marketing research, the bibliometric analysis includes a wide variety of terms and outputs from appropriate softwares like R studio. The objective philosophy is used in bibliometric analysis, which applies a quantitative inquiry method on writings (Faruk et al. 2021; Barrot 2021; Aria and Cuccurullo 2017). With the exception of earlier bibliometric studies, the scope of this study has been enlarged to encompass all scholarly resources and materials created between 1993 and 2022, such as articles, conference proceedings, books, chapters, etc. The study considers both quantitative and qualitative elements of digital marketing by retrieving the quantity of publications and the citation data, as well as an analysis of multiple factors such as authors, organizations, countries, keywords, sources, and paper productivity on an annual basis. The current study intends to examine and discover digital marketing publishing trends and patterns from 1993 to 2022 by examining the most productive countries, publications, and authors; significant journals; authorship patterns; collaboration networks; and worldwide collaborative trends.

14.2 Methodology

14.2.1 Method

The bibliometric analysis method was utilized to examine and discover publication patterns and developments in digital marketing literature. Bibliometrics is a quantitative/statistical tool for analyzing trends in the publication of research articles, conference papers, and other academic documents (Moed et al. 1985; Moral-Muñoz et al. 2020). It can be used to analyze a discipline's growth, publication trend lines, and patterns, and among its most important characteristics is its capacity to review published studies without approaching the scholars personally (Faruk et al. 2021; Saleem et al. 2021).

14.2.2 Database Selection

Web Of Science (WoS) knowledge repository was used for the present study because it is regarded as one of the most trustworthy, popular, powerful, and dependable knowledge databases in the world just like Scopus (Bahuguna et al. 2022; Liu et al. 2022).

14.2.3 Search Query

The relevant and most appropriate search queries were used to retrieve data. The keywords used were "Digital marketing" OR "Electronic Commerce Marketing".

14.2.4 Inclusion/Exclusion Criteria

The time frame considered in the study is for the research works published in Web of Science (WoS) from 1993 to 2022. All articles, review articles, proceedings, book chapters, and book reviews from business and management areas were filtered and included.

14.2.5 Data Analysis

After data was retrieved from WOS database. Data extraction and analysis were done with the help of software packages such as R studio, MS Excel, Biblioshiny, Mendeley reference manager etc.

14.3 Results and Discussion

14.3.1 Summary of Analysis

Table 14.1 presents the results of a bibliometric analysis of 987 digital marketing-related documents published between 1993 and 2022. The summary findings from the analysis are presented in Table 14.1.

Documents published in the WOS database during this period received an average of 4.37 citations per year. The higher average number of citations per document indicates a rapid hike in the quantum of scholarly papers in the field of digital marketing. The findings also revealed that 2226 distinct authors contributed to the field of digital marketing during this time period. Furthermore, single-author documents accounted for 153 of the totals.

The number of documents per author is 0.443, while the number of co-authors per document is 2.62. This indicates that a significant amount of research studies in the

Table 14.1 Summary statistics

Description	Result
Timespan	1993:2022
Documents	987
Average years from publication	9.71
Average citations per document	46.96
Average citations per year per doc	4.37
References	42,872
Authors	2226
Authors of single-authored documents	148
Authors of multi-authored documents	2078
Single-authored documents	153
Documents per author	0.443
Authors per document	2.26
Co-authors per documents	2.62
Collaboration index	2.49

Source Biblioshiny output

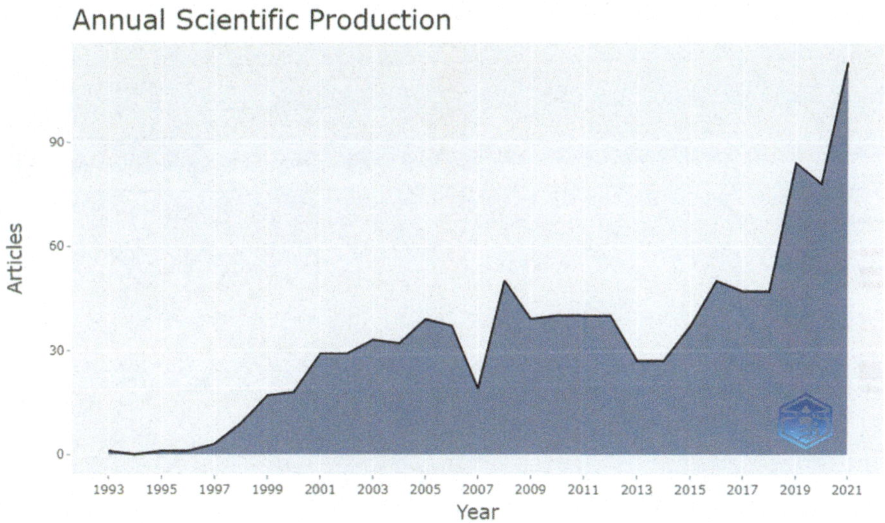

Fig. 14.1 Annual scientific production in digital marketing. *Source* Biblioshiny output

development of digital marketing are conducted in collaboration with other authors, as evidenced by the collaboration index of 2.49. However, a substantial number of single-author articles are also published.

14.3.2 Performance Analysis

The key trend in the annual scientific production in the field of digital marketing from 1993 onwards shows a positive upward growth trend. This growth is the result of the increasing number of internet and social media users. Figure 14.1 shows the annual scientific production in the field of digital marketing.

14.3.3 Association Between Sources, Authors, and Keywords

Figure 14.2 depicts the relationship between authors, keywords, and sources, with the left column containing the names of the authors, the middle column containing keywords, and the right column containing the journal name. This confirms that many of the authors chose electronic commerce as their keyword. However, commerce, electronic market, digital market, and many other closely related keywords with digital marketing have also been used in different research articles.

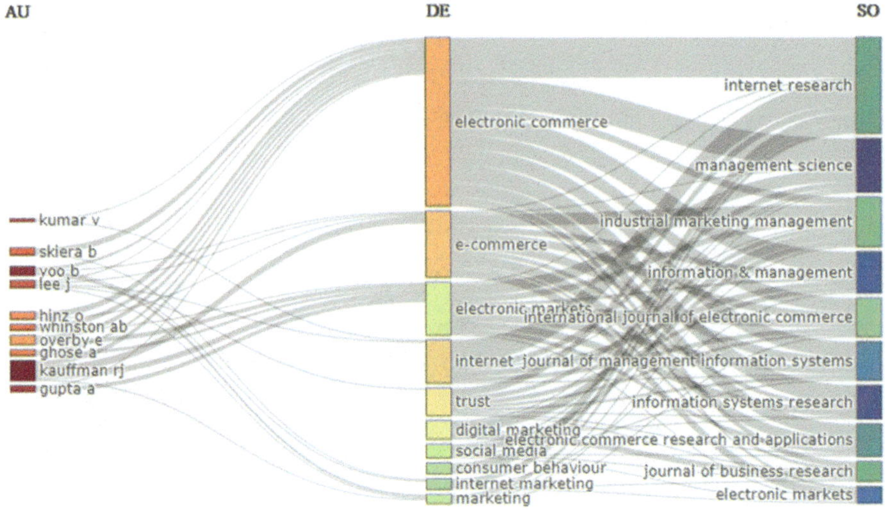

Fig. 14.2 Three-factor analysis of the relationship among author, keyword and source. *Source* Biblioshiny output

14.3.4 Analyzing the Contribution of Journals

In order to point out the most contributing journals, the most relevant sources category has been taken. It is found from Fig. 14.3 that the International Journal of electronic commerce had the highest contribution in this domain. This journal published 48 research papers within the specified period. Industrial marketing management with 42 documents, Journal of Business Research with 38, and Internet Research with 37 documents are the main contributors to the field of digital marketing.

14.3.5 Contribution and Collaborations by Countries

Among different countries' contributions in scholarly works on digital marketing, the bibliometric analysis found that the USA had contributed the most (770 papers) followed by China 277 papers, UK 191 papers. In case of India, only 31 papers were published in the field of business and management. Hence, there is a scope for future research in the field of digital marketing.

When the country collaboration map on digital marketing literature is considered, USA emerged as the top collaborator with China (45 publications), followed by USA and Canada (22 publications each), USA and UK (17 publications. When India is considered, it has collaborations with USA (nine publications) and with UK three publications. Norway, Qatar, Saudi Arabia, Sweden, Switzerland, Thailand, and Arab Emirates were the least active collaborators among all countries (Fig. 14.4).

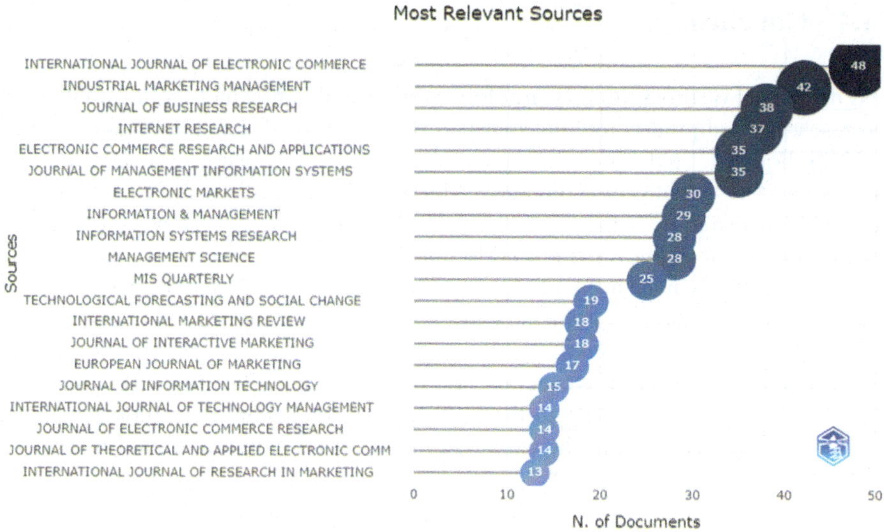

Fig. 14.3 Most relevant sources. *Source* Biblioshiny output

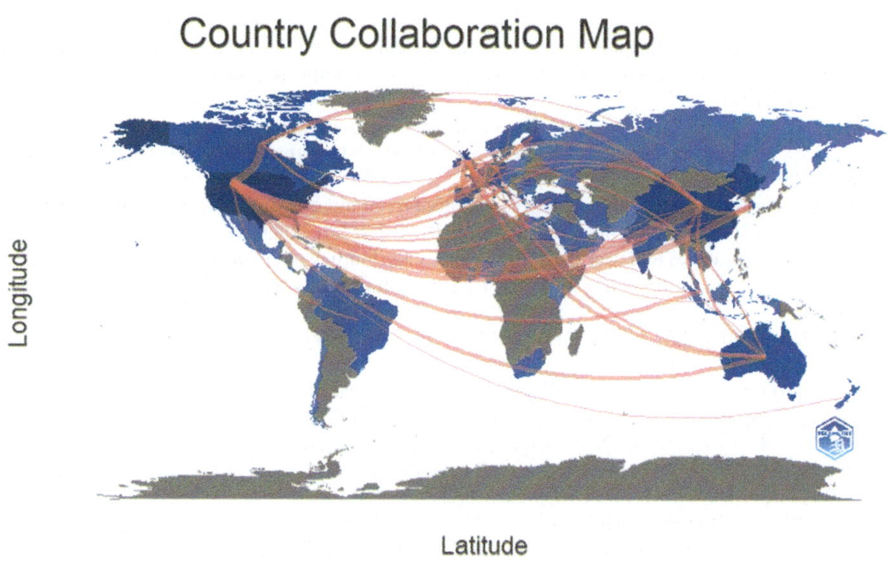

Fig. 14.4 Countries' scientific production and collaboration. *Source* Biblioshiny output

14.4 Conclusion

The main goal of this scholarly article was to provide an exhaustive evaluation of the research on digital marketing that had been published. For the period 1993 to 2022, bibliometric analysis on the topic of digital marketing using keywords "Digital marketing" OR "Electronic Commerce Marketing" was performed. This study can be used by researchers in the field of digital marketing to design strategies focusing on developing subjects (as evidenced by the data visualization in this study). They can also identify the most influential publications, authors, and journals in this field of research. By analyzing 987 published articles, conference papers, and book chapters from the Web of Science database, the study provided new insights in the field of digital marketing research. Academic interest in this area has gradually increased from 1993 to 2022, according to the findings. In conclusion, this study has made an important contribution to the expanding body of literature on digital marketing research.

References

Akhil MP (2022) Employing bibliometric analysis to identify emerging technologies in the insurance industry. In: Big data analyt insurance mark pp 207–220. https://doi.org/10.1108/978-1-80262-637-720221011

Aksnes DW, Langfeldt L, Wouters P (2019) Citations, citation indicators, and research quality: an overview of basic concepts and theories. SAGE Open 9(1). https://doi.org/10.1177/2158244019829575

Aria M, Cuccurullo C (2017) Bibliometrix: an R-tool for comprehensive science mapping analysis. J Informet 11(4):959–975. https://doi.org/10.1016/J.JOI.2017.08.007

Bahuguna PC, Srivastava R, Tiwari S (2022) Two-decade journey of green human resource management research: a bibliometric analysis. Benchmarking. https://doi.org/10.1108/BIJ-10-2021-0619

Bala and Verma 2018 Bala M, Verma D (2018) A critical review of digital. Int J Manag 8(10):321–339

Barrot JS (2021) Scientific mapping of social media in education: a decade of exponential growth. J Educ Comp Res 59(4):645–668. https://doi.org/10.1177/0735633120972010

Chaffey D, Ellis-Chadwick F (2019) Digital marketing. https://books.google.co.in/books?hl=en&lr=&id=-1yGDwAAQBAJ&oi=fnd&pg=PT17&dq=digital+marketing+research+paper&ots=XiWt7TjK32&sig=pfMkIhzvv3sQ9QsxMlaybcw26H4

Desai DMV (2019) Digital marketing: a review. Int J Trend Sci Res Dev Special Issue(Special Issue-FIIIIPM2019):196–200. https://doi.org/10.31142/IJTSRD23100

Faruk M, Rahman M, Hasan S (2021) How digital marketing evolved over time : a bibliometric analysis on scopus database Heliyon How digital marketing evolved over time : a bibliometric analysis on scopus database. Heliyon 7(December):e08603. https://doi.org/10.1016/j.heliyon.2021.e08603

Kraus S, Durst S, Ferreira JJ, Veiga P, Kailer N, Weinmann A (2022) Digital transformation in business and management research: an overview of the current status quo. Int J Inf Manage 63:102466. https://doi.org/10.1016/J.IJINFOMGT.2021.102466

Liu Z, Ren L, Xiao C, Zhang K, Demian P (2022) Virtual reality aided therapy towards health 4.0: a two-decade bibliometric analysis. Int J Environ Res Publ Health 19:1525. https://doi.org/10.3390/IJERPH19031525

Moed HF, Burger WJM, Frankfort JG, Van Raan AFJ (1985) The use of bibliometric data for the measurement of university research performance. Res Policy 14(3):131–149. https://doi.org/10.1016/0048-7333(85)90012-5

Moral-Muñoz JA, Herrera-Viedma E, Santisteban-Espejo A, Cobo MJ (2020) Software tools for conducting bibliometric analysis in science: an up-to-date review. Profes Inform 29(1):1–20. https://doi.org/10.3145/epi.2020.ene.03

Morris N (2009) Understanding digital marketing: marketing strategies for engaging the digital generation. J Direct Data Digit Mark Pract 10(4):384–387. https://doi.org/10.1057/dddmp.2009.7

Murillo-Ramos L, Huertas-Valdivia I, García-Muiña FE (2022) Exploring the cornerstones of green, sustainable and socially responsible human resource management. Int J Manpow. https://doi.org/10.1108/IJM-12-2021-0696

Parsons A, Zeisser M, Waitman R (1998) Organizing today for the digital marketing of tomorrow. J Interact Mark 12(1):31–46. https://doi.org/10.1002/(sici)1520-6653(199824)12:1%3c31::aid-dir4%3e3.0.co;2-x

Rose JJ, Biju AV, MP A (2020) Mergers and acquisitions-an analysis of citation and co-author networks using bibliometrix package. Int J Manag (IJM) 11(11):1383–1391. https://doi.org/10.34218/IJM.11.11.2020.130

Saleem F, Khattak A, Rehman SU (2021) Bibliometric analysis of green marketing research from 1977 to 2020 bibliometric analysis of green marketing research from 1977 to 2020. MDPI 9(1). https://doi.org/10.3390/publications9010001

Yasmin A, Tasneem S, Fatema K (2015) Effectiveness of digital marketing in the challenging age: an empirical study. J Int Bus Res Mark 1(5):69–80. https://doi.org/10.18775/IJMSBA.1849-5664-5419.2014.15.1006

Chapter 15
Impact of Digital Taxation for Sustainability Among Taxpayers With Special Reference to Bangalore

S. V. Malvika and Nikhila Vikram

Abstract Digitalization is rapidly changing all aspects of human life, our society and also our economy at large. The scale and speed of change brought about by digital transformation are remarkable and pose a variety of challenges for policymakers. E-filing is a prominent inclusion in the world of digitalization. Filing of tax with the help of digitalization in other words called E-filing is user-friendly and easier and also more efficient and accurate when compared to manual paper returns. It eliminates the need to make physical paper copies and attachments such as postage etc. Digitalization of tax is a good initiative taken by the government but people who are not well versed in computers, senior citizens or people from rural areas will find it difficult to file tax. Considering this as our problem statement we initiated a primary study to find out the opinion of taxpayers for which a questionnaire was circulated and based on the responses received by a few tax and non-taxpayers in and around Bangalore, an analysis has been brought down which gives answers to various questions amongst the main objective behind every question was to determine the impact of digitalization of tax on the taxpayers. This paper examines the attitude of taxpayers towards digitalization and the challenges and concerns faced by them. It further states the various ways how the tax payment can be made user-friendly.

Keywords Digital taxation · Sustainability · Taxpayers

15.1 Introduction

Tax is a word which is derived from the Latin word 'Taxare' or 'Taxo', the meaning of which is to assess the total value. Across the globe, every Government imposes taxes on its citizens for the various services provided by them. Taxes are imposed and taken by the state. It was in 1850 that Sir James Wilson formally introduced the

S. V. Malvika (✉) · N. Vikram
New Horizon College, Bengaluru, Karnataka, India
e-mail: malvikasv1998@gmail.com

A. Edward J. et al. (eds.), *Digital Transformation for Business Sustainability*,
Contributions to Environmental Sciences & Innovative Business Technology,
https://doi.org/10.1007/978-981-99-7058-2_15

tax in India. The Indian Income Tax Act of 1860 marks the watershed moment for taxation in India. It was through this act that centrally organized taxation began in India. The act was introduced to recover the losses the government suffered from the 1857 military mutiny.

The procedure by which we can transform raw information into a digital platform is called digitalization. The outcome of this procedure can be represented as object, image, sound, document, or signal by generating a line of numbers that describe a set of samples. This process of digitalization is rapidly changing all aspects of human life, our society and also our economy at large. The rate at which digitalization is growing is now notable in all our public policies.

It is also changing the nature of policy-making itself, through the emergence of a new range of tools to support the development and implementation of policies.

Digitalization in the Income tax Department started with the setting up of the Directorate of Income tax (Systems) in 1981. Electronic filing is the process of submitting tax returns over the internet using tax preparation software that has been preapproved by the relevant tax authority Initially computerization of processing of challans was taken up.

Filing of tax with the help of digitalization in other words called E-filing is user-friendly and easier and also more efficient and accurate when compared to manual paper returns. It eliminates the need to make physical paper copies and attachments such as postage etc. It also helps in easy and fast returns of refunds. This platform has paved the way for quick processing which was once a severe hurdle in our nation. Verification, vouching, filing and computations have all been worked efficiently and improved with the help of digitalization of taxing.

This paper contains a literature review for finding the research gap. It has highlighted the problem existing behind this research. The objectives of the study were set. The research methodology used was justified. The data analysis was followed by findings, suggestions and conclusion to this study. The limitations and the scope for further research were illustrated.

15.2 Review of Literature

According to a study made by Rosario and Chavali (2020), it throws light on the importance of digitalization in the current scenario and how it is compelling tax authorities to amend the tax rules and to make them more adaptable to digitalization. This paper further focuses on the need for advancement of tax administration and collection methods.

Ojo (2020) emphasizes on highlighting why present challenges may prove difficult to resolve given the multilateral dimension and importance of engaging a global consensus whilst also contributing to the literature by drawing attention to measures which could be adopted to address such challenges. Further, it highlights why addressing existing challenges are pivotal in the unlocking of greater benefits and potentials of digitalization.

Sarkar and Das (2018) deal with digitalization and its improvement regarding the efficiency and customer experience in several fields including the financial transaction areas. Digitalization has made a huge impact on the financial transactions which has led to gaining customer trust. This paper further emphasizes on other benefits of digitalization like better customer satisfaction, operational efficiencies etc.

The above research confirms that there has been no research conducted so far on this topic in Bengaluru.

15.3 Objectives of the Study

- To find out the relationship between age demography and ease of filing.
- To ascertain the level of difficulty and concerns involved in e-filing.

15.4 Research Methodology

The primary data for the study was collected by circulating an online questionnaire and the secondary data was collected through various news articles and information available. The study could get responses from 66 responses from in and around Bangalore, the majority of which are in the age group of 20 to 55. This study is performed to bring light to the perceptions of the people about e-filing. This study further emphasizes on the concerns faced by the people during e-filing. The study is empirical in nature.

15.5 Data Analysis and Interpretation

This section consists of two types of analyses. Percentage analysis, one-way ANOVA and chi-square test were used to test the data which was collected.

15.5.1 Percentage Analysis

The figures given below highlight the results obtained from the survey which was conducted.

Figure 15.1 shows the place of respondents 85.1% are from Bangalore urban, 5.9% from Bangalore rural and 9% are not from Bangalore.

Figure 15.2 shows the age groups of our respondents 50.7% belong to the age category 20–35, 40.3% represent 36–50 age category and 9% represent the 51 & above category.

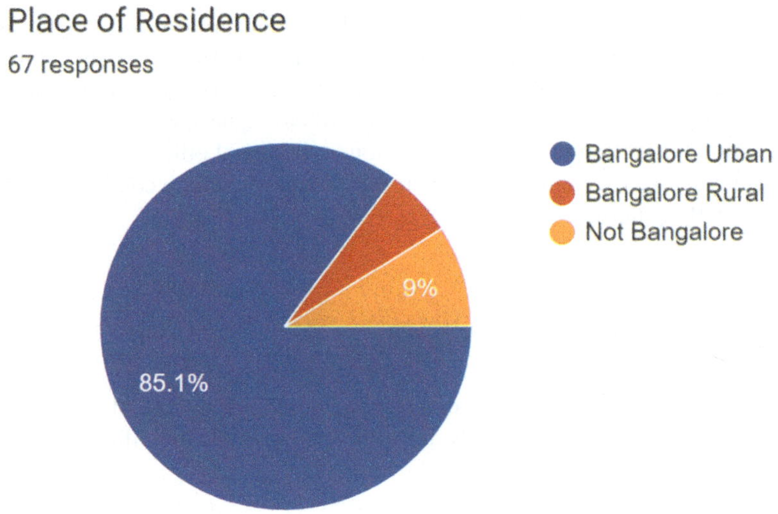

Fig. 15.1 The respondent's area

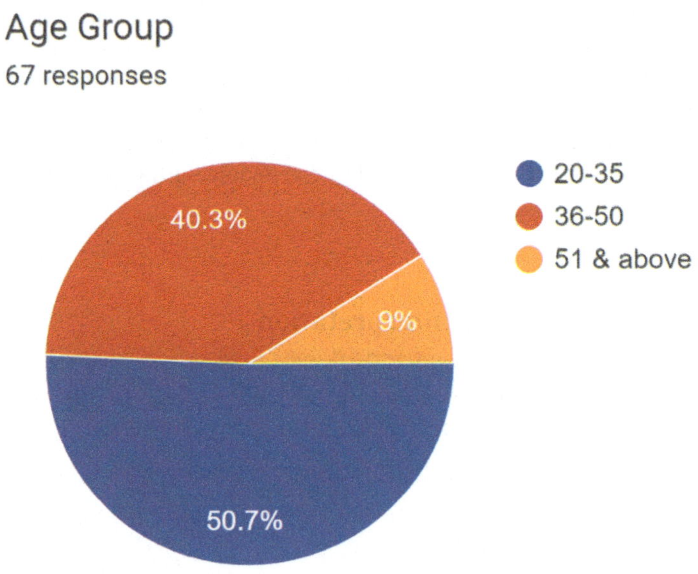

Fig. 15.2 The respondent's age

Figure 15.3 shows that 67.2% of the respondents are taxpayers.

Figure 15.4 shows that 76.1% of respondents prefer online mode of filing.

Figure 15.5 shows that 34% of respondents have filed tax for the first time.

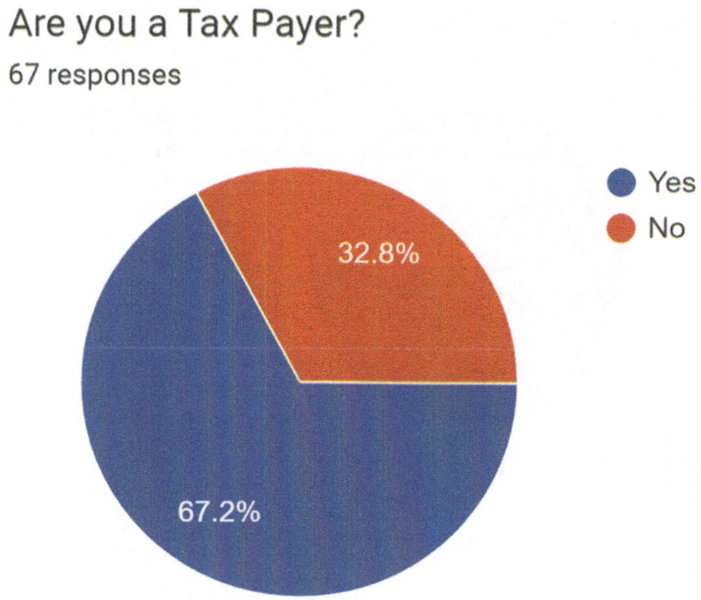

Fig. 15.3 The respondent's taxability

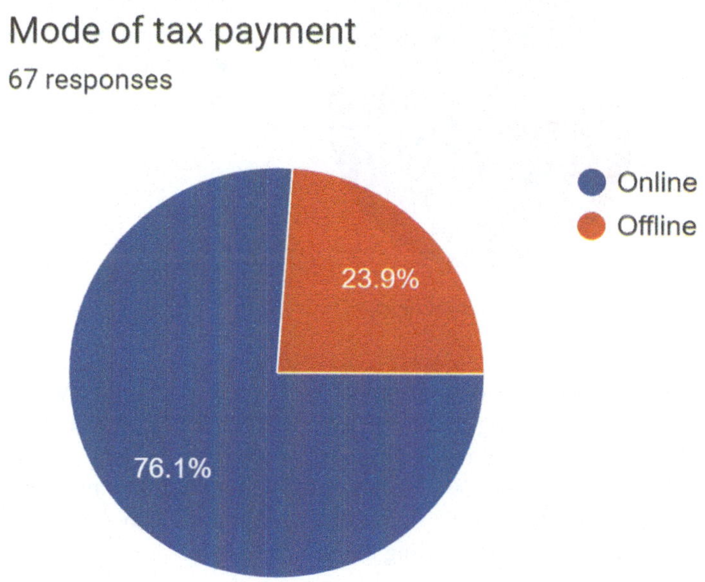

Fig. 15.4 The respondent's tax payment mode

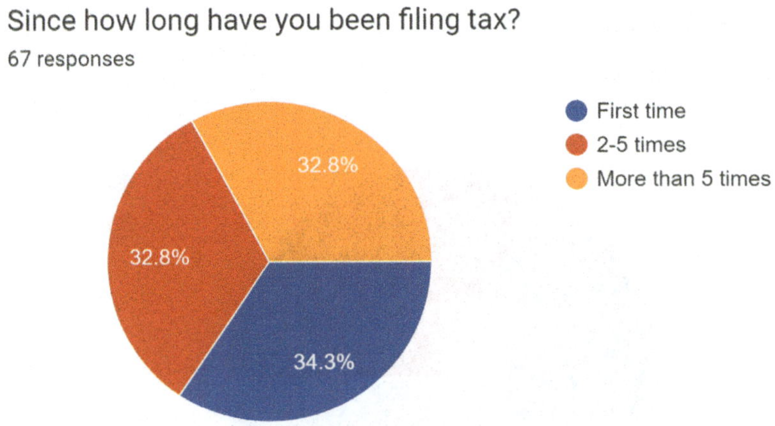

Fig. 15.5 The respondent's tax payment timeline

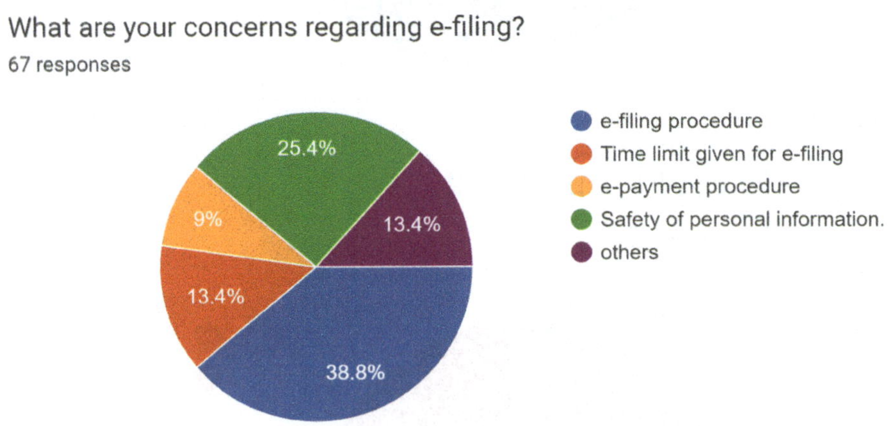

Fig. 15.6 The respondent's tax payment timeline

Figure 15.6 shows the various concerns faced by the taxpayers of which the most predominant one is the e-filing procedure carrying 38% following which is the safety of personal information weighing 25.4%.

Figure 15.7 shows that 76.1% of respondents preferred online mode of payment.

Figure 15.8 shows that 49.3% of respondents preferred to set up call centres, conduct workshops and set up of filing assistance units.

15.5.2 Chi-Square Test

The hypotheses used for the analysis are listed below:

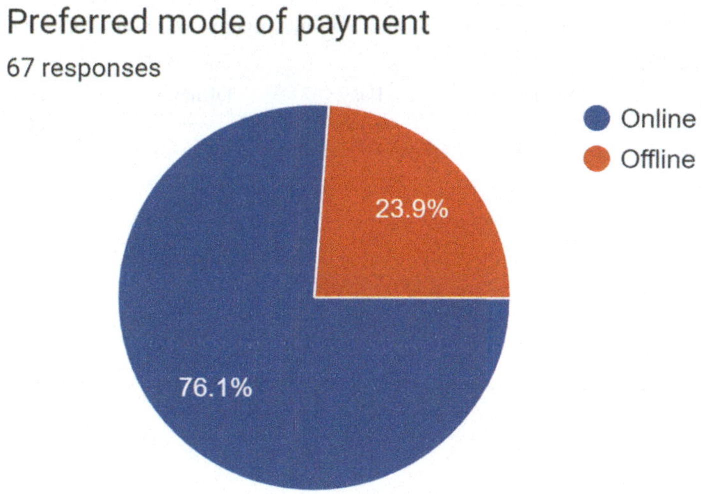

Fig. 15.7 The respondent's preferred mode of tax payment

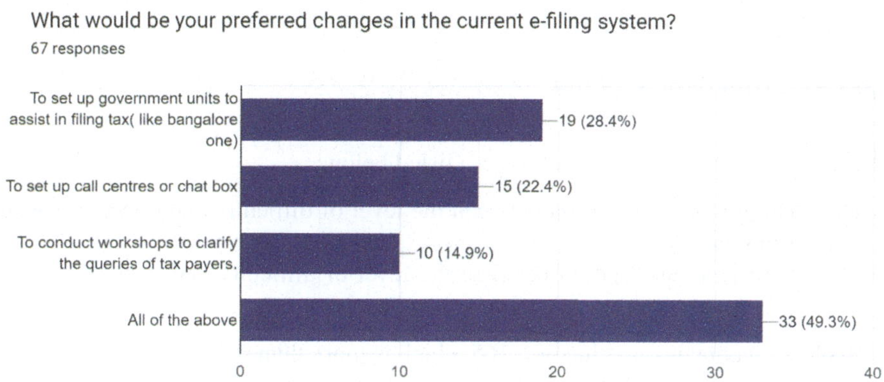

Fig. 15.8 The respondent's aspirations on tax filing reforms

Ho1: There is no significant difference between age and ease of filing.
H1: There is a significant difference between age and ease of filing (Table 15.1).

From Table 15.2 it is observed that the p-value is 0.311 which is greater than 0.05 which indicates that the alternate hypothesis can be rejected. Hence null hypothesis is accepted. There is no significant difference between age demographics and ease of filing tax.

Table 15.1 Response spread pertaining to ease of filing based on age of the respondents

Ease of filing*Age				
Age group	Moderate	Easy	Difficult	Total
20–35	23	9	2	34
36–50	15	8	4	27
51 & Above	2	2	2	6
Total	40	19	8	67

Table 15.2 Chi-square test results pertaining to age of the respondent and ease of filing

	Value	df	p-value
X^2	4.78	4	0.311
N	67		

Table 15.3 Response spread pertaining to level of difficulty and concerns faced while filing tax

	Sum of Squares	df	Mean square	F	Sig
Between groups	1.315	3	9.43	1.79	0.165

15.5.3 One-Way ANOVA

The hypotheses used for the analysis are listed below:

Ho2: There is no association between the level of difficulty and concerns faced while filing tax.
H2: There is an association between the level of difficulty and concerns faced while filing tax.

Table 15.3 contains p-value of 0.165 which is more than 0.05. Hence, there is no association between the level of difficulty and the concerns involved in e-filing.

15.6 Conclusion

The primary and secondary sources have revealed that people prefer to file tax online but they face some issues. Therefore, as part of the study, the suggestions are given below:

- To set up government units to provide assistance in the tax filing procedures
- The above suggestion will also provide employment opportunities to people and help those unable to file because of a lack of digital knowledge
- To set up call centres or provide a chat box facility to fix issues during the process of filing.

- To educate the younger generation through workshops and hands-on training in the aspects of filing tax.

The study is limited to the city of Bangalore. The study is limited to a period of two weeks. Future research can be undertaken on other demographic variables.

References

Ojo ROB (2020) Tax challenges arising from digitalization. Research Gate. https://www.researchgate.net/publication/340077800_Tax_Challenges_Arising_from_Digitalization

Rosario S, Chavali LDK (2020) Digitization of taxation in the changing business environment & base erosion & profit shifting (BEPS) special reference to India. Eur Sci J 16(1):61–74

Sarkar R, Das S (2018) Digitalization and its impact on financial transactions in India. Frontiers of Digital Finance

Chapter 16
A Study on Price Discovery of Jeera and Mustard Seeds Sustainable Trading in NCDEX

Madhu Druva Kumar and Cherian Thomas

Abstract India being an agricultural economy, agricultural commodities play a very important role. The commodity derivative market was started a way long back in 1874 in India but it regained its importance only from 2000 and within the span of a short time it has gained its importance in price discovery, hedging and arbitrage process. An efficient commodity futures market attracts a number of investors in the futures market. At this juncture, there are many complications related to price signals, volatility spillovers and price discovery in the spot and futures of Indian commodity market. The present study examines the price discovery of agricultural commodities traded in the NCDEX with reference to the spot prices and future prices of jeera and mustard seeds for the period of January 2020 to December 2021. The study investigates the integration between spot and future price through Johnson integration test and the causal relationship whether they have a bidirectional or unidirectional relationship by using Granger Causality Test.

Keywords Commodity derivative market · Price discovery

16.1 Introduction

The agriculture sector is the backbone of an economy which provides the basic ingredients to mankind and now raw materials for industrialization. Agricultural commodity markets have occupied a very important place in the economic growth and progress of countries. The future market trading in commodities evolved in the middle of the nineteenth century and in India, the first commodity exchange was started in 1875 with Bombay cotton trade exchange and after Independence there was a ban on commodity future trading and only in 2000 did the commodity derivative market was restarted with a National Exchange (Nirmala et al. 2015). The lead–lag

M. D. Kumar (✉) · C. Thomas
Kristu Jayanti College, (Autonomous), Bengaluru, Karnataka, India
e-mail: madhumalathi@kristujayanti.com

A. Edward J. et al. (eds.), *Digital Transformation for Business Sustainability*, Contributions to Environmental Sciences & Innovative Business Technology,
https://doi.org/10.1007/978-981-99-7058-2_16

study between spot and futures prices has shown inconclusive and mixed results. Some of the researchers have concluded that spot price plays an efficient role. The spot prices can lead to the futures price if the convenience yield is high enough (Meng yuan, Ocean University of China 2015). While others have concluded that future price is more efficient and leads to the discovery of subsequent spot price. Major economic fiction of the futures market is price discovery.

16.1.1 Mustard Seeds

Mustard seeds are usually grown in India during the Rabi season and it is also called as Rape, Oilseed rape, Rapeseeds, and Oilseed rape. After soya bean oil, Mustard seed is the third leading source of vegetable oil in the world and it is the world's second-leading source of protein meal (Vasantha and Mallikarjunappa 2015).

16.1.2 Jeera

Cumin is a flowering plant from the family Apiaceous. Cumin seed is used in perfumery and as a flavoring agent in a variety of cuisines. In India, Jeera is sown from October to November and harvested in February. The plant thrives on rich, well-drained sandy loam soil with daytime temperatures of around 30 degrees. Cumin crop takes about 110–115 days to reach maturity. Unjha in Gujarat is the main trading centre for Jeera in the country.

Review of Literature

Ahuja (2006) concluded that Commodity Derivatives market has grown enormously in terms of number of commodities, volume and value of commodities. This paper shows the working of NCDEX and MCX and the various commodities traded in the exchange.

Kaur and Anjum (2013) described the Indian commodity futures market. Several commodity indices are fake, yet they do show a general trend. The major goal of this research is to ascertain if prices in the Indian commodities futures market accurately represent the functioning of the market, given that it is less developed than financial derivatives markets and is constrained by India's chaotic history of policy reversals.

Nirmala and Swarna (2017) examine the available notional price indices in the commodity market and conclude that multi-commodity indices that include metals and energy products and that have a clear and efficient price dissemination in national and international markets behave similarly to equity indices in terms of efficiency and information flow. On the other hand, these traits appear at more muted levels in agricultural indexes.

Madhavi (2017), this paper examine the price discovery process and the effectiveness of hedging and risk minimization in the selected commodities and found that there is a unidirectional relationship where the price is discovered in the future market and manages the risk very efficiently and also helps the hedger to know the factors improving hedging effectiveness..

Samal (2017), the paper evaluates the efficiency of Indian cotton futures prices for the years 2013 to 2015. ADF test is used to check the stationarity. The results show that both the variables are stationary at the level. The VAR model suggests that there is a cointegration and the causality test has further indicated that futures markets does not influence the spot prices for cotton.

(Palaniappan et al. 2016). Studied the cointegration relationship and causal relationship of corn, wheat, soybeans, soybean meal and oil, feeder and live cattle, as well as lean hogs to test whether they are cointegrated or not and whether they have bidirectional relation or unidirectional relation. It was found the prices are mostly discovered in spot market and all the commodities have long-run cointegration.

Objectives

- To check stationarity of spot and future prices of selected commodities
- To estimate the long-run and short-run relationship between spot and future prices.
- To check the unidirectional or bidirectional causality between spot and futures prices

Hypothesis

- H_0: Time series data is non-stationary.
- H_0: There is no long-run cointegration between Spot Price and Future Price for Jeera and Mustard Seeds agricultural products.
- H_0: There is no Unidirectional/Bidirectional causality between spot and future prices.

Data Collection and Methodology

Source of data: This study is based on the secondary data, which is collected from NATIONAL COMMODITY AND DERIVATIVE EXCHANGE LIMITED (NCDEX). And the study is conducted based on two agricultural commodities Mustard seed and Jeera. The commodities selected for the study are selected on the basis of their trading frequency on NCDEX for a period of 2 years of January 2020 to December 2021.

1. **Tools and Techniques**
    - Descriptive Statistics
    - Augmented Dickey-Fuller test
    - Johnson Cointegration Test
    - Granger Causality Test
2. **Analysis**

Tables 16.1 and 16.2 show values of Skewness for both Jeera and Mustard [spot price & future price] indicating data almost exhibit a normal distribution and Kurtosis

Table 16.1 Descriptive statistics of Jeera spot and future commodity prices

	Jeera price	
	Spot	Future
Mean	14,126.35	14,123.27
Median	13,929.48	13,932.50
Maximum	16,290.0	16,575.0
Minimum	12,863.65	12,535.0
Std. dev	790.60	832.24
Skewness	1.353	1.035
Kurtosis	4.326	3.773
Jarque–Bera	180.94	97.36
Probability	0.000	0.000
Sum	6,752,395	6,750,925
N	478	478

Source Author compilation

implying that the distribution are peaked relative to normal for both Jeera and Mustard. The Jarque–Bera test indicates that none of the commodity markets is normally distributed.

Unit Root Test result for data stationarity:

Hypothesis:

H_0: Time series data is non-stationary.
H_1: Time series data is stationary.

Table 16.2 Descriptive statistics of Mustard spot and future commodity prices

	Mustard price	
	Spot	Future
Mean	6026.36	5968.17
Median	5772.50	5518.00
Maximum	8805.00	8822.00
Minimum	3836.0	3836.00
Std. dev	1448.78	1481.03
Skewness	0.277	0.392
Kurtosis	1.867	1.844
Jarque–Bera	33.13	40.65
Probability	0.000	0.000
Sum	3,013,180	2,984,088
N	500	500

Source Author compilation

In order to detect the order of integration we employed both the Augmented Dickey-Fuller (1979) and Phillips-Perron (1988) unit root tests, whose results are shown in Tables 16.3 and 16.4 for the first difference. The null hypothesis of a unit root is not rejected for all indices in the log level, whereas it is rejected when they are taken in their log first differences. It emerged from the preliminary ADF results that all the four variables (Spot and Future Jeera commodity, Spot and Future Mustard commodity) were not stationary at levels [I(0)]. However, all four variables were subjected to the first difference and found that they are stationary (Table 16.5 and 16.6) in the first difference [I(1)], thus satisfying the necessary conditions for cointegration. Thus, H_1 is accepted that data is stationary at first difference.

Johansen Cointegration test:

Null hypothesis: There is no long-run cointegration between spot price and future price for Jeera and Mustard agricultural product.

Alternative Hypothesis: there is a long-run cointegration between spot price and future price for Jeera and Mustard agricultural product.

First, with respect to Jeera commodity, from the cointegration results depicted in Table 16.5, it is observed that the p-value is 0.000 and 0.023 under None& At most 1 respectively. This shows that there is one cointegration equation between spot and

Table 16.3 Results of unit root tests at first difference of Jeera

		Jeera	
		Spot	Future
Augmented Dickey-Fuller test statistic (ADF)	t-statistic	−19.728(0.000*)	−17.274(0.000*)
	p-value		
	1% level	−3.443	−3.443
	5% level	−2.867	−2.867
	10% level	−2.569	−2.569

Source Author compilation

Table 16.4 Results of unit root tests at first difference of Mustard

		Mustard	
		Spot	
Augmented Dickey-Fuller test statistic (ADF)	t-statistic	−21.408(0.000*)	−23.649(0.000*)
	p-value		
	1% level	−3.443	−3.443
	5% level	−2.867	−2.867
	10% level	−2.569	−2.569

Source Author compilation
* Significance at 1% level. P-values are provided in the parenthesis

Table 16.5 Johansen cointegration test result for spot price and future prices of Jeera Commodity

Hypothesized no. of CE (s)	Eigen value	Trace statistics	0.05 critical value	Prob **	Max-Eigen value
None*	0.0662	37.588	15.494	0.000	32.418
At most 1*	0.0108	5.178	3.841	0.023	5.170

Trace Test Indicates 2 cointegrating eqn(s) at the 0.05 level
*denotes rejection of the hypothesis at the 0.05 level
**Mackinnon-Haug-Michelis(1999) p-values
Source Author Compilation

Table 16.6 Johansen cointegration test result for spot price and future prices of Mustard

Hyptosized no. of CE (s)	Trace statistics	Prob**	Prob**
None*	10.169	0.268	0.218
At most 1*	0.261	0.609	0.609

Trace Test Indicates 2 cointegrating eqn(s) at the 0.05 level
*denotes rejection of the hypothesis at the 0.05 level
**Mackinnon-Haug-Michelis(1999) p-values
Source Author Compilation

future prices of Jeera. On the contrary, the p-value of Mustard is 0.268 (at None) and 0.609 (almost 1) indicating that there is no cointegration equation between spot and future price of Mustard commodity. Nonetheless, the Max–eigen value test indicates there exists only cointegrating equation in Tables 16.5 and 16.6 respectively. Since, two variables that are cointegrated, would on average, not drift apart over time, this concept provides insight into the long-run relationship between the two variables (Spot price & Future price) and testing for the cointegration between two variables.

The Johansen cointegration test result for spot and future prices (Table 16.6) for Mustard shows that p-value is (0.218) which is greater than 0.05 and hence there exist no cointegration equations between spot and future prices.

Granger Causality for Jeera and Mustard commodity:

H_0: There is no Unidirectional/Bidirectional causality between spot and future prices.
H_1: There is Unidirectional/Bidirectional causality between spot and future prices.

The granger causality test result (Table 16.7) shows that there is unidirectional causality from Future Prices to Spot Prices with respect to Jeera commodity. In a sense, it emerges that Future Prices influence the changes in the Spot Prices for Jeera. On the other, there is no causality relationship with respect to Mustard commodity.

Findings

- The median future price for Mustard is Rs 5772.50 with a peak price of Rs 8822 and with a lowest value of Rs 3836 during the sample time period. The mean of JEERA spot price is Rs 14,126 (see Table 4.3) and Rs 14,123 for future prices

Table 16.7 Granger causality for Jeera and Mustard commodity

Null hypothesis	Obs	F-statistic	Prob	Remarks
Jeera				
D_1_SP_Jeer does not granger cause D_1_FP_JEER	476	0.620	0.431	Accept
D_1_FP_MUST does not granger cause D_1_SP_JEER		36.772	0.000*	Reject
Mustard				
D_1_SP_MUST does not granger cause _1_FP_MUST	498	3.755	0.0532	Accept
D_1_FP_MUST does not granger cause _1_SP_MUST		0.871	0.350	Accept

Source Author compilation

with a standard deviation of Rs 790.6 (spot) and Rs 832.24 (future) respectively. The median spot price is Rs 13,929 with a peak price of Rs 16,290 and with a lowest value of 913.85 for Jeera

- The null hypothesis of a unit root is not rejected for all indices in log level, whereas it is rejected when they are taken in their log first differences. It emerged from the preliminary ADF results that all the four variables (Spot and Future Jeera commodity, Spot and Future Mustard commodity) were not stationary at levels [I(0)]. However, all four variables were subjected to the first difference and found that they are stationary (Table 4.5) in the first difference [I(1)], thus satisfying the necessary conditions for cointegration. Thus, H_1 is accepted that data is stationary at first difference
- The cointegration results show that there is one cointegration equation between spot and future prices of Jeera and Mustard
- The Granger causality test result shows that there is unidirectional causality from Future Prices to Spot Prices with respect to Jeera and Mustard

16.2 Conclusion

The study was undertaken to investigate the price discovery of selected agricultural commodities which are Mustard seed and Jeera. It is concluded that there is a long-run association and future market influence spot market in the price discovery of Mustard seed and jeera. The result of Granger causality between futures and spot price returns suggested that the futures market dominates the spot market of both the commodities.

16.3 Scope for Further Research

This Study has scope for further research to determine the effectiveness of arbitrage between futures and spot prices and also using macroeconomic variables to forecast volatility can be studied and we can also undertake a comparative study on Indian commodity markets with worldwide markets.

References

Ahuja NL (2006) Commodity derivatives market in India: development, regulation and futures prospective. Int Res J Finance Econ 1(2)

Kaur HP, Anjum DB (2013) Commodity derivatives market in India. Int Res J Bus Manag IRJBM 5. ISSN No: 2322–083X

Madhavi A, Chalam GV (2017) Performance evaluation of commodity derivatives exchanges in india. Int J Res Commerce Econ Manag 7(3). ISSN 2231–4245

Mengyuan Q (2015) An empirical study on the relationship between corn futures prices of China and the United States. J Int Stud 8(3):191. https://doi.org/10.14254/2071-8330.2015/8-3/15

Nirmala K and Swarna C (2017) Examining the relationship between spot and future price of crude Oil, Kaav Int J Econ Commerce Bus Manag 4:309–314. ISSN 2348–4969

Nirmala K, Munilakshmi R, Sandhya V (2015) Price discovery in commodity markets: a study of Indian cardamom market in multi commodity. Indian J Account XLVII (1) ISSN-0972–1479

Palaniappan Shanmugam V, Raghu RR (2016) Relationship between spot and futures markets of selected agricultural commodities in India. J Bus Financ Aff 5:1. https://doi.org/10.4172/2167-0234.1000160

Samal GP (2017) Price discovery efficiency of cotton futures market in India. Agric Econ Res

Vasantha G and Mallikarjunappa T (2015) Lead-lag relationship and price discovery in indian commodity derivatives and spot market: an example of pepper. IUP J Appl Finan **21**(1)

Chapter 17
Rational Influence of Debt Structure Patterns on Sustainable Profitability: A Pragmatic Evidence from Tata Steel

T. K. Murugesan and S. Vidhya

Abstract The primary crux of this empirical study is to analyze the rational influence of debt structure patterns on the profitability position of Tata Steel. The financial data of Tata Steel were collected for the period of ten years, i.e., from 2013 to 2022. Three econometric models were developed by the researchers to analyze the influence of debt content patterns on the profitability position of the sample company. For measuring the profitability of the company, the financial indicator of ROTA was employed and STD to TA, LTD to TA and TD to TA were employed in this empirical study to identify the debt structure patterns of the company. To pragmatically analyze the cogent association between debt structure patterns and profitability position, three significant accounting models were developed and analyzed by using Ordinary Least Square (OLS) indicators. Since three accounting models clearly showed that the overall debt structure patterns have the substantial stimulus on the profitability position of Tata Steel. The best accounting model was also selected based upon the OLS indicators such as OLS AIC criterion, OLS SCH criterion and OLS HQ criterion.

Keywords Debt structure · Profitability · ROTA · STD to TA · LTD to TA and TD to TA · OLS estimators

17.1 Preamble

The debt structure patterns of the company depend on the major financing decisions of the company. Financial policy is an imperative verdict that every firm needs to make a point because it not only influences the firm's estimated cash flows but also the firm's profitability and liquidity position in the future and makes a clear-cut association between debt content and profitability position of the firm (Baker et al.

T. K. Murugesan (✉) · S. Vidhya
Kristu Jayanti College (Autonomous), Bengaluru, India
e-mail: murugesan@kristujayanti.com

A. Edward J. et al. (eds.), *Digital Transformation for Business Sustainability*, Contributions to Environmental Sciences & Innovative Business Technology, https://doi.org/10.1007/978-981-99-7058-2_17

2011). There are many sources available for funding long-term capital expenditure projects (Gleason et al. 2000). The authors indicated that there are many advantages of raising funds using debt instruments because the company can enjoy tax deductibility on the interest payments.

Most of the manufacturing firms employed debt structure as a key indicator in measuring and analyzing the financial health of the firm (Salim and Yadav 2012). The studies on debt structure patterns on the financial position of the company, the researchers have done a study to determine how financial resources of a company are effectively used to enhance the profitability position of the company (Gleason et al. 2000). For analyzing that, they have used significant debt structure patterns and profitability position of the company and in turn the valuation of shareholder's wealth (Salim and Yadav 2012). Debt structure measures the major debt patterns and frameworks of the sample firm. Whenever there is a turbulent and fluctuating economic condition, the company can raise funds through issuing debt instruments. In case of bad economic conditions, the debt structure of the company will have a negative influence on the profitability position of the company (Herciu and Ogrean 2017).

17.2 Review of Literature

The main objective of a corporate firm is to maximize the valuation of the shareholders. Many studies have been conducted to assess the impact of capital structure on profitability position of the company (Joshua 2015). Corporate firms generally focus on maximizing the wealth and valuation of equity shareholders through the drastic reduction in the overall cost of capital (OCC) and enhancing the market prices of the financial instruments (Amatya 2020). One of the ways of minimizing the overall cost of capital is to finance the corporate firm's capital through a proper mixture of debt content and equity content (DeMarzo and Fishman 2017). Corporate firms normally finance their business through the issue of debt or equity instruments (Chaklader and Chawla 2016). Many factors are to be taken into consideration while deciding the optimum mixture of debt and equity and most of the corporate firms find it a challenge to optimize the composition of the debt content and equity content in the process of increasing the value of the firm (Herciu and Ogrean 2017).

The pragmatic study was conducted to determine the significant positive relationship between the ratios of long-term and short-term debt solvency patterns and their effective influence on the profitability position of the sample company (Abor 2015). However, it is clear that there was a negative association between the major indicators of STDS & LTDS ("Short-Term Debt Structure" and "Long-Term Debt Structure") (Baker et al. 2011). It is also clear from the study that the STDS & LTDS have a significant influence on the Return on Capital Employed (ROCE). With respect to the rational relationship between various financial indicators of the corporate firm, many researchers studied the STDS & LTDS and its pragmatic impact on financial viability and effective performance of corporate firms (Baum et al. 2006).

17.3 Statement of the Problem

In order to achieve and sustain an optimum debt structure for the firm is one of the significant priorities in the turbulent business environment. In a couple of recent years, several empirical studies were conducted on the influence of capital structure patterns on the profitability position and pecuniary performance of the corporate firms. The desirable mixture of debt and equity is essential for maximizing shareholders' wealth. There was limited research on the impact of debt structure on the profitability position of the company. No study was found in the area of debt structure and its impact on the profitability position of the company by developing suitable accounting models. This paper aims to close this gap by providing some insights into the cogent relationship between debt structure and its influence on the profitability of Tata Steel.

17.4 Main Crux of This Pragmatic Study

(1) To pragmatically analyze the debt structure of Tata Steel by applying suitable financial indicators.
(2) To determine the Return on Total Assets (ROTA) as a measure of profitability of the sample company.
(3) To examine pragmatic association between debt content structure and profitability position of Tata Steel

17.5 Research Framework

The methodology of this research study is empirical in nature and the nitty–gritty of this research is to determine the cogent relationship between debt structure and profitability position of Tata Steel for the period of 2013 to 2022. This study also explores the degree to which the STD to TA, LTD to TA and TD to TA have an empirical effect on the profitability. Three financial indicators of STD to TA, LTD to TA and TD to TA were considered explanatory variables of this pragmatic study and the ROTA (Return On Total Asset) was taken as regressand variable. The capital structure of the sample company has become ubiquitous on the basis of profitability indicators of the company. For pragmatically achieving the primary objective of this study, the accounting framework was developed and modeled to identify the cogent relationship between regressand and regressors of this accounting model (Fig. 17.1). This accounting framework is deemed to be a direct linear model that designates the significant relationships to test the proposed hypotheses of this study.

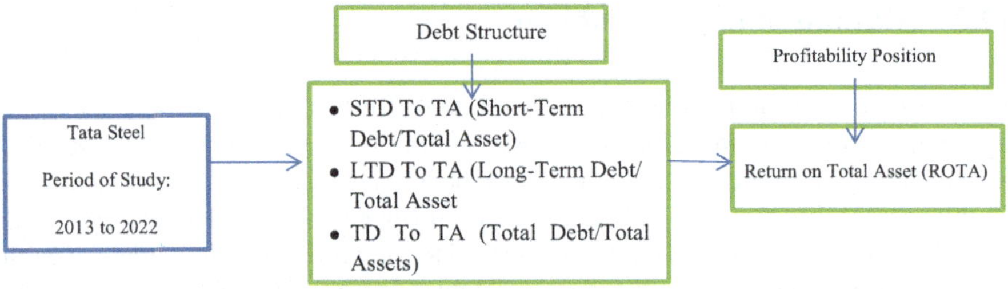

Fig. 17.1 Statistic relationship between debt structure and ROTA

17.6 Hypotheses of the Study

In this study, the following hypotheses were framed to figure out a cogent relationship between three indicators of debt content structure and one indicator of the profitability position.

- H_1: ROTA is significantly influenced by STDTA.
- H_2: ROTA is significantly influenced by LTDTA.
- H_3: ROTA is significantly influenced by TDTA.

17.7 Model Estimation:

Model 1: In model 1, the STD to TA (or STDTA) is taken as regressor and Return On Total Assets (ROTA) is taken as regressand. Symbolically the model is formed as follows:

$$ROTA = \beta_0 + \beta_1 STDTA + e_i$$

Model 2: In model 2, the STD to TA (or STDTA) and the LTD to TA (or LTDTA) are taken as regressors and Return On Total Assets (ROTA) is taken as regressand. Symbolically the model is formed as.

follows:

$$ROTA = \beta_0 + \beta_1 STDTA + \beta_2 LTDTA + e_i$$

Model 3: In model 3, Short-Term Debt to Total Asset (STDTA), Long-Term Debt to Total Asset (LTDTA) and Long-Term Debt (LTD) are taken as regressors and Return On Total Assets (ROTA) is taken as regressand. Symbolically the model is formed as follows:

$$ROTA = \beta_0 + \beta_1 STDTA + \beta_2 LTDTA + \beta_3 TDTA + e_i$$

17.8 Comparison of Three Estimated Models:

The outcome of the OLS estimators of the three models are summarized in Table 17.1. Three models are compared on the basis of the coefficient of Regressand, coefficient of Regressors, the R^2 statistical value, F-statistic value, Prob (F-statistic), AIC, SCHC and HQC.

The R^2 value of 0.909 is very high for Model 3 as compared to Model 1 and Model 2. The greater the statistical value of R^2, the better the accounting model developed in this research that the regressors have an influence on the regressand. The value of F-statistic is found to be very high in Model 3 and the Model 3 is highly significant in explaining the regressand (i.e., ROTA). Since the p-value (F-statistic) of Model 2 and Model 3 is less than 0.05 (the significance level taken for the study), we can safely reject the null hypothesis of this pragmatic study and determine that debt content structure has a substantial stimulus on the profitability position of Tata Steel.

The AIC ***(Akaike Info Criterion), SCHC (Schwarz Criterion) and (HQC) Hannan-Quinn Criterion*** were generally employed to make a significant choice of the accounting models which are competing each other in terms of analyzing between regresand and regressors. The higher the statistical values of these OLS indicators, the worse the accounting model is predicting the study variable. In our study, Model 3 is the best model as the values of these criteria are the lowest as compared to statistical values of important criteria developed in Model 1 and Model 2. The lesser the statistical values of OLS criteria, the better the fitness of the accounting model developed in this research.

Managerial Implications

In this study, three econometric models were developed by academic researchers to analyze the influence of debt content structure patterns on the profitability position of Sample Company. For measuring debt structure and its rational influence on profitability position of the sample company, the accounting indicators of STDS and LTDS were employed in this empirical study. The outcome of this study clearly

Table 17.1 Comparison of three estimated models

Particulars	Model 1	Model 2	Model 3
Regressand	ROTA	ROTA	ROTA
Regressors	STDTA	STDTA & LTDTA	STDTA, LTDTA & TDTA
R-squared	0.053755	0.842566	0.909455
F-statistic	0.454467	18.73153	20.08839
Prob (F-statistic)	0.519222	0.001548	0.001568
Akaike info criterion	5.630923	4.037428	3.684271
Schwarz criterion	5.691440	4.128204	3.805305
Hannan-Quinn criterion	5.564536	3.937847	3.551497

Source Author compilation

indicated that there was a cogent association between the debt structure patterns and profitability position of the sample company. Since three accounting models clearly exhibited that the overall debt structure pattern has a pragmatic influence on the profitability of Tata Steel. In our study, Model 3 is found to be the best model based upon the OLS indicators such as OLS AIC, OLS SCHC and OLS HQC statistical estimators. It clearly indicated that the STDTA, LTDTA and TDTA have a pragmatic influence on Tata Steel's profitability position during the period of this study. Future studies can be conducted to include other accounting and financial indicators to determine debt structure patterns of companies in different sectors on the sustainable profitability.

References

Abor J (2015) The effect of capital structure on profitability: an empirical analysis of listed firms in Ghana. J Risk Finance 6:438–45. London

Amatya K (2020) Determinants of capital structure in Canadian non-financial firms: a recent study. Toronto Metropolitan University, Toronto, Canada

Baker HK, Veit ET, Powell GE (2011) Factors influencing dividend policy decisions of Nasdaq firms. Financ Rev 3:19–38. Canada

Baum CF, Schafer D, Talavera O (2006) The effects of short-term liabilities on profitability: the case of Germany. In: Money Macro and Finance (MMF) research group conference 2006 61. Money Macro and Finance Research Group

Chaklader B, Chawla D (2016) A study of determinants of capital structure through panel data analysis of firms listed in NSE CNX 500

DeMarzo PM, Fishman MJ (2017) Optimal long-term financial contracting. Rev Financ Stud 20:15–32

Gleason KC, Mathur LK, Mathur I (2000) The interrelationship between culture, capital structure, and performance: evidence from European retailers. J Bus Re 2:185–191. Toronto

Herciu M, Ogrean C (2017) Does capital structure influence company profitability? Stud Bus Econ 12:50–62

Salim M, Yadav R (2012) Capital structure and firm performance: Evidence from Malaysian listed companies. Proc-Soc Behav Sci 65:156–166. London

Chapter 18
Synthesis of Factors Impacting Usage of OTT Platforms: Sustainability of OTT Industry Post Pandemic

A. Pushpa

Abstract India's OTT streaming market has made a remarkable growth and prominence, positioning it as one of the largest and fastest-growing in the world. With a diverse range of popular streaming platforms such as Voot, Alt Balaji, Zee5, MX Player, Amazon Prime, Disney + Hotstar, Netflix, Sony Liv, Aha, and numerous others. The main objective is to Synthesise factors impacting the usage of OTT platforms. The study adopted a cross-sectional descriptive approach. Three hundred and eighty responses were collected using convenience sampling method. Six factors, i.e., content, convenience, features, price & quality, satisfaction and willingness to continue usage of OTT platforms were considered. The effect of these five factors along with willingness to continue usage is tested on subscription decisions using correlation and regression analysis. Given the highly diverse nature of the Indian OTT service market, this descriptive study offers valuable insights into streaming service providers operating in this fiercely competitive landscape. The findings can assist OTT service providers in making informed decisions and devising effective strategies to cater to the specific needs and preferences of Indian consumers.

Keywords OTT platforms · Satisfaction · Willingness to continue

18.1 Introduction

After the boom of the internet, there was a drastic change in the news media and general entertainment sector. Constant technological change has fortified the development and growth of OTT platforms in the past decades. The BCG marketing research firm in its industry report "Entertainment goes online" claimed entry and uptake of OTT platforms has rehabilitated the media content consumption. Indian entertainment market is complex in terms of media content consumption since the

A. Pushpa (✉)
Kristu Jayanti College (Autonomous), Bengaluru, India
e-mail: apushpakn@gmail.com

A. Edward J. et al. (eds.), *Digital Transformation for Business Sustainability*, Contributions to Environmental Sciences & Innovative Business Technology,
https://doi.org/10.1007/978-981-99-7058-2_18

transition from Doordarshan to DTH (Direct-to-home) to OTT (Over-the-top) platforms has undeniably evolved. The rapid growth in digital content consumption is the result of easy access to smart devices and affordable data plans. Statista projects the OTT segment revenues to reach $3.76bn in 2023 and the content market to peg $5bn in size and is expected to grow at an annual growth rate of 10.01% by 2027. According to the Global Web Index report, India's streaming market is one of the biggest and fastest-growing in the world with biggies like Voot, Alt Balaji, Zee5, MX Player, Amazon Prime, Disney + Hotstar, Netflix, Sony Liv, Aha and many more. According to a report published by IAMAI and Kantar Research, India's internet users are expected to reach 900 million by 2025, from ~622 million internet users in 2020, increasing at a CAGR of 45% until 2025. Studies reveal a promising and massive rise of OTT platforms and its usage in India.

During the two years of the pandemic (2020 and 2021), OTT (or streaming) category dominated headlines as were shut for the most part and television content remained stagnant, OTT emerged as the new and exciting destination for quality entertainment content. OTT made its debut in 2008 with BIGFlix launched by Reliance Entertainment followed by Amazon Prime and Netflix in 2012. The period saw a spate of films released online directly, as also the launch of new platforms in different Indian languages. Established platforms such as Disney + Hotstar, Prime Video, Netflix and the likes, upped the momentum, rolling out new series by the week. Over-the-Top (OTT) services provide digital users with movies and shows without the need for cable or satellite television. The media and entertainment industry has become the leading player in the OTT market due to the rising number of online video consumers. Various reports indicate the increasing market and consumer demand for personalized content on OTT platforms. One positive outcome of this trend is that, with cinema screens being inaccessible for a prolonged period, audiences have grown fond of OTT video streaming and are unlikely to abandon it anytime soon. As a result, OTT platforms will remain the top preference for individuals seeking entertainment at the click of a button. The satisfaction of viewers is paramount in the success of a streaming platform, and they tend to favor platforms with simple and user-friendly navigation. While content is crucial, user experience (UX) plays an equally significant role in the success of any OTT platform. The majority of users prefer a streamlined interface that is easy to navigate and allows them to find the desired content with minimal effort. Cluttered or complex interfaces tend to be less appealing to users.

The emergence and adoption of Over-the-Top (OTT) services have revolutionized content accessibility and empowered consumers by eliminating the need for cable, broadcast, and satellite television platforms that have traditionally controlled or distributed such content. This B2C approach has become a key differentiator of the platform and medium. According to industry experts and analysts, regional OTT platforms already account for approximately 40–45 percent of the overall OTT consumption. The latest FICCI-EY report predicts that by 2025, regional language content consumption on OTT platforms will surpass 50 percent of the total market. This highlights the high potential for the future of this industry. OTT TV has enabled the democratization of quality content to small audiences, with reduced technology and delivery costs and unrestricted distribution space. Consequently, an increasing number of targeted services are emerging, not only for hobbyists or communities but also for enterprises looking to engage with their employees and customers.

OTT platforms have gained popularity among the younger generation and a wider audience due to the diverse range of content, which not only entertains but also informs. The surge in video streaming platforms can be attributed to discounted subscription plans that offer unlimited content, as well as a growing number of documentary-style films that present information in an engaging manner. The widespread usage of smartphones and internet availability has also contributed to the growth of these platforms. One of the most popular formats on OTT is web series, which has become a hot trend.

Consumers' motives for embracing or resisting this innovation can coexist over time and may be attributed to similar factors. The objective of this study is to examine the factors driving the adoption and resistance to OTT services in India, while also exploring the demographic characteristics of the users. With the backdrop of changed consumption behavior among viewers, the present study aims to comprehend the pertinent factors influencing the willingness to continue and subscribe to steaming services in the mere future. The study makes a novel contribution to the literature by evaluating the factors affecting willingness to continue subscriptions and also the moderating role of satisfaction on willingness to continue. The subsequent section delves into the pertinent literature related to the study, followed by a section on the methodology. After presenting a comprehensive explanation of the study's design and approach in the methodology section, the study proceeds with sections dedicated to data analysis, results, findings, and discussion, as well as implications and conclusion.

18.2 An Overview of Study Constructs: Review of Literature

Digital illustrations have changed the way of lives and the way media is consumed, (Nair 2021). The entertainment industry is never been static, it has witnessed a change in the form and format of entertainment. OTT has emerged as a dominant industry in the media and entertainment sector due to the increase in online video consumers. Viewers use new and varied technologies to watch movies, documentaries, web series, and news, using personal gadgets. As a result, well known OTT companies like Amazon Prime, ALT Balaji, Disney + Hotstar, Jio TV, Netflix, and Zee5 provide video content, movies etc. through internet (Sahu and Kumar 2021).

OTT refers to video streaming media services that are good substitutes for traditional distribution services through cable, DTC and Cinema Halls. The media services are directly offered to viewers that can be accessed on electronic gadgets using internet; it could be a paid subscription, free, ad-supported or pay to watch streams that are offered including iTunes and video on demand. In today's entertainment work, OTT has become the buzzword signifying entertainment anywhere and everywhere. This new digital era has undeniably opened up avenues for streaming services to connect and engage with customers in real-time. Earlier studies by scholars have highlighted the impact of digital services on entertainment. Though the adoption of OTT services is a widespread phenomenon and has brought radical transformations in the entertainment industry, it has received recent attention of scholars (Sharma and Kumar 2020).

Literature on digital and online contexts provides an extensive list of factors that comprise ease of use, usefulness, quality, attitude, trust, risk, security, customer engagement, service experience, habit, satisfaction, content quality, cost, Content variety, Privacy, quality of service and many more. Also, customers have numerous OTT services to choose from, attracting and engaging customers is a great deal for the service providers (Singh and Yadav 2020).

From the aforesaid discussion, in order to remain competitive, it becomes imperative for the service providers to focus on key drivers that impact the customer's decision to stay connected to the firm and provide it with a steady flow of sales in future, Sood and Kumar 2021). The present study has focused on the following precursors impacting the willingness to continue the usage of OTT platforms. Each of the variables considered is reviewed below (Han et al. 2019).

Pricing

Consumers tend to prefer platforms that offer affordable subscription plans or pay-per-view options for premium content (Sharma and Kumar 2020; Sood and Kumar 2021). The impact of price on the adoption, usage, and satisfaction of OTT platforms has been extensively studied in recent literature. Several studies have found that price is a critical factor that influences consumers' decision to subscribe to OTT platforms (Lee and Hong 2020). Price has a significant impact on the frequency of usage and satisfaction levels of users. Nair and Gupta (2020) found that consumers who paid a lower subscription fee were less likely to use the platform frequently and had lower satisfaction levels than those who paid a higher subscription fee. Lee and Hong (2020) found users' perception of fairness in pricing significantly influenced their satisfaction levels. Moreover, several studies have examined the impact of pricing strategies on OTT platform usage and willingness to pay. O'Reilly and Mirabito (2018) found that dynamic pricing strategies, such as offering discounts for longer-term subscriptions or bundling services, were effective in increasing user retention and willingness to pay. Overall, the literature suggests that price is a critical factor that impacts the adoption, usage, and satisfaction of OTT platforms. OTT platform providers need to carefully consider their pricing strategies to attract and retain users while also ensuring that the pricing is perceived as fair by their users.

Content

Park and Kim (2018) found that the perceived quality of content, including its relevance, variety, and uniqueness, was positively related to user satisfaction with the OTT platform. The type of content available on the OTT platform has been found to be a significant factor in attracting and retaining users. Park and Kim (2018) users are more likely to use OTT platforms that had a diverse range of content, including movies, TV shows, and original content. Sharma and Kumar (2020) users prefer OTT platforms that offer a mix of regional and international content. Furthermore, the quality of content, including its production value, has been found to influence user satisfaction. They found that users were more likely to be satisfied with OTT platforms that provided high-quality video and audio content. They perceived the quality of content, including its production value, positively related to user satisfaction. Literature suggests that the content available on OTT platforms plays a significant role in attracting and retaining users. OTT platforms that provide a diverse range of high-quality, personalized content are more likely to satisfy users and encourage continued usage.

Quality

One of the primary factors that influence the usage of OTT platforms is the quality of content available on them. Consumers tend to prefer platforms that offer a diverse range of high-quality content that caters to their interests and preferences. There has been a considerable amount of research on the influence of content quality on the usage, willingness, and satisfaction of OTT platforms. Many studies have found that the quality of content has a significant impact on the user's decision to subscribe to an OTT service and their overall satisfaction with the platform. Yang and Shen (2019) found that the perceived quality of content positively influenced user satisfaction and intention to use OTT services. The quality of content significantly affected user satisfaction with the platform and their intention to continue using it. Wei and Zhang (2020) found that the perceived quality of content positively influenced user willingness to pay for OTT services. They found that content variety and quality were significant predictors of OTT adoption among Indian consumers. Overall, these studies suggest that the quality of content is a critical factor in determining the success of an OTT platform. OTT providers must focus on delivering high-quality content to attract and retain users, and in turn, improve their willingness and satisfaction to use the platform.

Features

Studies have explored the impact of OTT platform features on user satisfaction and willingness to continue using the service. One study found that platform features such as personalized recommendations, user-friendly interface, and high-quality video streaming significantly influenced user satisfaction and retention (Sharma and Kumar 2020). Another study found that features such as easy navigation, user control over content, and content availability played a critical role in user satisfaction and willingness to continue using the service. They found that features such as ease of use, content discovery, content recommendation, and personalized user experience significantly impact user satisfaction and willingness to continue using OTT platforms. OTT platforms with features such as user interface, video quality, search options, and content recommendation have a positive impact on user satisfaction and willingness to continue using the platform. (Sharma and Kumar 2020). It was found that features such as seamless streaming, high-quality content, personalized user experience, and easy navigation significantly influence user satisfaction and willingness to continue using OTT platforms. They revealed that features such as variety of content, user-friendly interface, personalized recommendations, and multiple device compatibility significantly impact user satisfaction and willingness to continue using OTT platforms. (Sharma and Kumar 2020) found that features such as user interface design, content recommendation, and social interaction significantly impact user satisfaction and willingness to continue using OTT platforms.

Convenience

Research has found that the ease and convenience of accessing content through OTT platforms positively affect user satisfaction and usage intentions. Singh and Yadav (2020) found that the convenience of OTT platforms, including anytime access to content, multiple device compatibility, and personalized recommendations, significantly impacted user satisfaction and willingness to continue using the platform. Similarly, a study by (Sharma and Kumar 2020) found that the convenience of OTT platforms, including easy navigation and search options, positively influenced user satisfaction and usage intentions. Kim and Sung (2019) found that the convenience of OTT platforms, including user interface design, content recommendation, and social interaction, positively influenced user satisfaction and usage intentions. Overall, the convenience of accessing and navigating content on OTT platforms is an important factor in user satisfaction and usage intentions. As such, OTT providers should focus on providing convenient and user-friendly features to attract and retain users. The ease of use and overall user experience of an OTT platform also play a significant role in its usage. Consumers tend to prefer platforms that are user-friendly, intuitive, and offer a seamless viewing experience across multiple devices (Sharma and Kumar 2020).

18.3 Statement of the Problem and Objectives

Investigating the factors impacting the usage of OTT platforms post-COVID-19 is crucial due to significant changes in consumer behavior during the pandemic. Lockdowns and social distancing measures have led people to rely heavily on digital platforms for entertainment. Understanding the sustained demand for OTT platforms and the factors influencing their usage is valuable for content creators, streaming platforms, and marketers. Factors such as convenience, content availability, pricing models, and user experience play a vital role in consumer satisfaction and loyalty. By studying these factors, stakeholders can make informed decisions on content creation, platform design, pricing, and marketing strategies. This research is essential for industry professionals and policymakers to adapt to the evolving media and entertainment landscape and provide quality digital entertainment experiences in the future.

The research paper aimed to address two main questions:

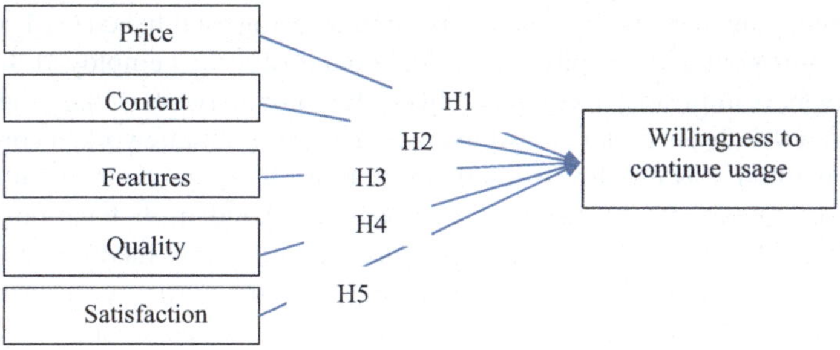

Fig. 18.1 Research framework. *Source* author compilation

1. What is the definition and classification of over-the-top (OTT) platforms within the media literature?
2. What are the factors that drive the adoption of OTT streaming media platforms?

By examining relevant literature, the researcher sought to gain a comprehensive understanding of the concept of OTT and its various classifications as described in media literature. This exploration aimed to provide clarity on the definition and categorization of OTT platforms. Furthermore, the researcher aimed to identify and analyze the factors that contribute to the adoption of OTT streaming media platforms. These factors could include technological advancements, content availability, pricing models, user experience, convenience, and other relevant aspects. By studying these factors, the research aimed to shed light on the key drivers behind the adoption of OTT platforms and their growing popularity among consumers.

18.4 Research Framework and Hypothesis

A research framework was proposed and the five hypothesis was framed. Figure 18.1 depicts the proposed framework and hypothesis outlined to measure the motivating factors that contribute to willingness to continue usage of OTT platforms post-pandemic. Figure 18.1 shows the Research Framework for the study.

18.5 Research Methodology

This research aims to shed light on the significance of OTT platforms and their impact on human consumption patterns. By examining the context of OTT platforms, we can gain a deeper understanding of how they have established themselves in the realm of interactive content. Through this study, we can gain valuable insights into the growth of OTT platforms and their subsequent dominance over other conventional media forms. The primary objective of this study was to assess the motivations and factors influencing viewers' inclination to continue using established OTT platforms. To achieve this goal, a descriptive cross-sectional design was employed. In order to obtain a diverse sample that accurately reflects the characteristics of the general population. Given the absence of an existing customer sample framework, a convenience sampling method was employed for this research. A comprehensive online questionnaire, designed with careful structuring, was circulated through various channels such as WhatsApp, e-mail-based groups, forums, and social media platforms. All respondents were required to answer all the questions in the survey, ensuring that no questionnaires were excluded from the analysis. The collected data was organized and stored in Excel spreadsheets for later importation into SPSS 22 for testing and analysis purposes. The survey items were measured using a five-point Likert scale, ranging from 1 (strongly disagree) to 5 (strongly agree).

18.5.1 Demographic Analysis

A descriptive analysis was conducted to gather information on the demographic profiles of the respondents. Analysis of demographic profiles provides insights into the characteristics of the respondents in terms of gender, age, education level, and income. Understanding these profiles can help contextualize the findings and interpret the results of the study. The findings reveal gender distribution Out of the total 350 respondents, 42% were male, while 58% were female. In terms of age distribution significant difference exists, majority of the respondents (86 percent) were less than 23 years of age, indicating a relatively young sample. Further, in terms of Education Level majority of the respondents were well-educated, with 62% holding a Bachelor's degree and 5% having a Master's degree. Approximately 33% had a Diploma or a lower level of education. In terms of income, 74% of the respondents had a disposable income of less than 50,000 per month. Additionally, 12% had a disposable income between 50,000 and 1 lakh, while 33% had a disposable income above 1 lakh.

18.6 Data Analysis and Interpretation

The statements utilized in the research were derived from prior studies, with each factor being measured using multiple items. Table 18.1 shows the Cronbach's Alpha value of each variable considered in the study which indicates that the questionnaire is considered reliable, as the value surpasses the minimum acceptable value of 0.7.

Table 18.2, signposts the reliability coefficient of each factor considered in the study is greater than 0.7 indicating that each variable has a high level of reliability and can be considered as a consistent and stable measure of the underlying construct it represents.

The KMO test assumption of having a value greater than 0.6 has been met with a value of 0.846, indicating that the sample is adequate up to 84.6%. Additionally,

Table 18.1 Reliability of each factor considered in the study

Factor	No. of statements	Cronbach's alpha
CONV	4	0.904
QLTY	2	0.807
CONT	4	0.741
FEAT	4	0.742
PRIC	2	0.827
SAT	3	0.909
WTC	3	0.915

Table 18.2 KMO and Bartlett's test

Kaiser–Meyer–Olkin measure of sampling adequacy		0.846
Bartlett's test of sphericity	Approx. Chi-square	325.938
	df	21
	Sig.	0.000

the significant value, which should be less than 0.05, has yielded a p-value of 0.000, indicating that the result is appropriate and normality has been achieved.

18.6.1 Correlation Analysis

A correlation analysis was conducted to understand the relationship between the independent variables convenience, quality, content, features, price, and dependent variables satisfaction, and willingness to continue usage of OTT platforms.

The results shown in Table 18.3 reveal that convenience has a significant positive correlation with content (r = 0.470**) and features (r = 0.679**), indicating that as convenience increases, so does the perceived quality of content and features. Price also has a significant positive correlation with content (r = 0.465**) and features (r = 0.662**), indicating that as price increases, so does the perceived quality of content and features. Quality has a significant positive correlation with content (r = 0.457**) and a weak positive correlation with features (r = 0.289*), indicating that as the perceived quality of the platform increases, so does the perceived quality of content.

Further, the study divulges that satisfaction has a strong positive correlation with convenience (r = 0.811**), content (r = 0.409**), features (r = 0.686**), and price (r = 0.569**), indicating that users who are more satisfied with the platform perceive it to be more convenient, with higher quality content, features, and price. And willingness to continue using the platform has a strong positive correlation with convenience (r = 0.851**), indicating that users who find the platform more convenient are more likely to continue using it. It also has a positive correlation with content (r = 0.369**)

Table 18.3 Pearson correlation with willingness to continue usage of OTT platforms

Variable	Pearson coefficient	Significance
Price	0.037	0.760
Content	0.470**	0.000
Features	0.679**	0.000
Quality	0.526**	0.000
Satisfaction	0.811**	0.000
WTC	0.851**	0.000

**Correlation is significant at the 0.01 level (2-tailed)

and features (r = 0.645**), indicating that users who perceive higher quality content and features are also more likely to continue using the platform.

18.6.2 Regression Analysis

A regression analysis was conducted to examine the relationship between a dependent variable and one or more independent variables. In the context of the study, the dependent variable is "willingness to continue usage," and the independent variables are "Satisfaction, Quality, Price, Content, Convenience, and Features". The goal is to understand how these independent variables influence the dependent variable.

With regression analysis, the study assessed the strength and significance of the relationship between the independent variables and willingness to continue usage. The analysis can help determine which independent variables have a significant impact on the dependent variable and to what extent.

The model summary is shown in Table 18.4 provides an overview of the regression model that was performed using the given predictors. The R-value of 0.894 indicates a strong positive correlation between the dependent variable and the set of predictors. The R-squared value of 0.799 suggests that approximately 79.9% of the variance in the dependent variable can be explained by the predictors in the model.

The adjusted R-squared value of 0.779 takes into account the number of predictors in the model and provides a more accurate measure of the model's goodness of fit. The standard error of the estimate, which is 0.448, represents the average distance that the observed values deviate from the predicted values. The predictors used in the model

Table 18.4 Model summary

Model	R	R square	Adjusted R square	Std. error of the estimate
1	0.894[a]	0.799	0.779	0.448

[a]Predictors: (Constant), SAT, QLTY, PRIC, CONT, CONV, FEAT

are CONVENIENCE, QUALITY, CONTENT, FEATURES, PRICE, and SATISFACTION. The constant term is included by default and represents the predicted value of the dependent variable when all of the predictors are equal to zero.

Table 18.5 shows ANOVA that the regression model is significant ($p < 0.05$) since the F-value of 41.003 is significant at the 0.000 level. The regression model explains a significant amount of the variance in the dependent variable, as evidenced by the high F-value and low p-value.

Table 18.6 shows the coefficients of each independent variable and their significance levels. The significant predictors are convenience, satisfaction, and price, indicated by their low p-values of 0.000, 0.000, and 0.400 respectively. The beta coefficients suggest that convenience and satisfaction have the strongest relationship with willingness to continue usage of OTT platforms. Specifically, for every unit increase in convenience and satisfaction, willingness to continue usage of OTT platforms increases by 0.497 and 0.400 units respectively. Price has a positive but weaker relationship with willingness to continue usage of OTT platforms, with every unit increase in price resulting in a 0.066 unit increase in willingness to continue usage of OTT platforms. Quality, content, and features are not significant predictors in the model as indicated by their high p-values.

The regression equation derived from the coefficients is:

Table 18.5 ANOVA a

Model		Sum of squares	df	Mean square	F	Sig.
1	Regression	49.392	6	8.232	41.003	0.000b
	Residual	12.447	62	0.201		
	Total	61.839	68			

[a]Dependent Variable: WTC
[b]Predictors: (Constant), SAT, QLTY, PRIC, CONT, CONV, FEAT

Table 18.6 Coefficients a

Model		Unstandardized coefficients		Standardized coefficients	t	Sig.
		B	Std. error	Beta		
1	(Constant)	0.107	0.321		0.332	0.741
	CONV	0.548	0.113	0.497	4.848	0.000
	QLTY	−0.010	0.056	−0.012	−0.178	0.859
	CONT	−0.124	0.114	−0.093	−1.089	0.281
	FEAT	0.062	0.119	0.057	0.525	0.601
	PRIC	0.061	0.073	0.066	0.847	0.400
	SAT	0.422	0.114	0.400	3.720	0.000

[a]Dependent Variable: WTC

$$\begin{aligned}\text{Willingness to continue OTT platforms} = {} & 0.107 + (0.548 \text{ x Convenience}) \\ & - (0.010 \text{ x Quality}) - (0.124 \text{ x Content}) \\ & + (0.062 \text{ x Features}) + (0.061 \text{ x Price}) \\ & + (0.422 \text{ x Satisfaction})\end{aligned}$$

18.7 Conclusion and Implications

The OTT platform market is experiencing rapid growth and can be considered as an evolved version of traditional television. Therefore, it can be inferred that OTT service providers can enhance their strategies by considering the aforementioned factors. It is crucial to acknowledge the significance of all the factors mentioned above and incorporate them into the process of building customer satisfaction and loyalty. The results suggest that convenience plays a crucial role in users' satisfaction and willingness to continue using OTT platforms, along with the perceived quality of content, features, and price. Therefore, OTT platforms should focus on improving convenience and offering high-quality content, features, and affordable pricing to increase user satisfaction and retention.

Overall, the model summary table indicates that the set of predictors used in the model have a strong positive correlation with the dependent variable and can explain a significant amount of the variance in the dependent variable. However, it is important to note that correlation does not imply causation and further research would be needed to establish any causal relationships between the variables. The study's limitations include a potentially limited and specific sample size, affecting generalizability. The cross-sectional design provides only a snapshot of user perceptions at one point in time, warranting the need for longitudinal studies. Self-reported data may be influenced by response bias or inaccuracies. Missing variables, like customer support and platform reliability, could impact user satisfaction and retention. Future research suggestions involve conducting comparative studies among different OTT platforms to identify best practices. Qualitative analysis, longitudinal studies, and exploring cultural influences would provide richer insights. Investigating user experience design and content diversity's impact on satisfaction could help optimize OTT platforms. Addressing these aspects will enhance understanding of customer satisfaction and retention in the evolving OTT market.

References

Han J, Kim W, Jeong E (2019) Factors affecting the intention to pay for online content among Millennials. Telematics Inform 37:135–145

Kim H, Sung Y (2019) Exploring the effects of OTT service quality on user satisfaction and the intention to use: focusing on the mediating role of user experience. Telematics Inform 36:101345. https://doi.org/10.1016/j.tele.2019.101345
Lee S, Hong S (2020) Effects of perceived fairness on user satisfaction and repurchase intention in OTT services: focusing on the mediating effect of trust. Telematics Inform 52:101417
Nair AR (2021) Impact of OTT platforms on viewing experience. J Res Bus Manag 9(8):18–30 (2021) ISSN(Online):2347–3002 www.questjournals.org
Nair S, Gupta A (2020) Determinants of users' satisfaction and loyalty towards video streaming platforms. J Retail Consum Serv 57:102159
O'Reilly N, Mirabito M (2018) Beyond the subscription model: new pricing strategies for video content platforms. J Bus Res 88:220–228
Park HJ, Kim H (2018) Investigating the determinants of video streaming services: a comparison of Netflix and Naver. Telematics Inform 35(4):1070–1081
Sahu N, Kumar R (2021) An empirical study on factors affecting consumer's adoption of OTT platform services in India. J Retail Consum Serv 58:102312
Sharma N, Kumar V (2020) Factors affecting the adoption of OTT platforms in India: an empirical study. J Indian Bus Res 12(4):384–405
Singh RK, Yadav SS (2020) Factors influencing adoption of OTT platforms in India: an empirical investigation. Int J Bus Emerg Mark 12(1):47–66
Sood A, Kumar A (2021) Factors influencing consumer adoption of over-the-top (OTT) platforms in India. J Retail Consum Serv 62:102603
Wei Y, Zhang L (2020) The influence of perceived quality, price, and social influence on willingness to pay for online video services. Int J Inform Manag 52:101992
Yang KC, Shen YC (2019) Exploring the relationships among video content quality, perceived value, satisfaction, and behavioral intention in the context of OTT video services. Telematics Inform 44:101270

Chapter 19
Digital Transformation for Sustainability in Bengaluru: Is It Happening in a Smarter Way?

Cherian Thomas and Madhu Druvakumar

Abstract Tier one cities are naturally assumed to be the hubs of digitalization. A closer look at rankings proves otherwise. In such a case, it is worthwhile to understand the problems faced by a city like Bengaluru. The study found that the nature of problems is systemic. It calls for action from the concerned seats of power. The study is an eye opener for the government and calls for imminent attention on a city that has potential to flourish leaps and bounds. This study specifically focuses on digitalization or digital transformation. Digitalization of daily use services leads to mass usage of such services. Therefore, it is totally a government centered initiative when it comes to the source of implementation of digital transformation in smart cities.

Keywords Digitalization · Bengaluru · Smart city · Ranking · National smart cities mission

19.1 Introduction

Bengaluru was the base for the Britishers before Independence. Later, it became a pensioner's paradise. The Information Technology boom has led to its present shape after 1999. The mere presence of information technology based companies does not make a city into a smart and digitalized one. This was proven true when Bengaluru was ranked 17th in India in terms of being Smart City as per National Smart Cities Mission in 2022. There are a wide variety of problems plaguing Bengaluru. This has made the city stoop so low in ranking despite being a state capital, Silicon Valley and start-up friendly city in the country.

National Smart Cities Mission defines a smart city as one which has infra and services that cater to the ease of living in a city. Among the different services

---

C. Thomas · M. Druvakumar (✉)
Kristu Jayanti College, (Autonomous), Bengaluru, Karnataka, India
e-mail: madhumalathi@kristujayanti.com

A. Edward J. et al. (eds.), *Digital Transformation for Business Sustainability*, Contributions to Environmental Sciences & Innovative Business Technology,
https://doi.org/10.1007/978-981-99-7058-2_19

and infra, this study specifically focuses on digitalization or digital transformation. Digitalization of daily use services leads to mass usage of such services.

Digital transformation is a slow but planned process. It does not occur naturally but it is affected by causes like natural calamities or the pandemic.

The stakeholder of a smart city with digitalization is cut across various demographic factors and is not restricted to a particular group. Therefore, its benefits are widespread. A major factor that plays a crucial role for this digital transformation is capital. This capital investment can only be done by a behemoth like a government. Therefore, it is totally a government centered initiative when it comes to the source of implementation of digital transformation in smart cities.

The article contains literature review section that contains past literature that had stressed on the topic. The discussion part delves on the possible solution to the problem. The conclusion part stresses on the need to digitalize Bengaluru. Additionally it comprises of limitations and the future scope of research.

19.2 Literature Review

Vallicelli (2018) confirmed that it is the need of the hour for every city to be smart since the concept of workplace had changed. In the knowledge economy, it is not restricted to a building. It is rather much beyond that. A public place can also turn into a workplace. In such a context, it is imperative that smart cities must embrace digitalization.

Hämäläinen (2020) studied the digitalization aspect in terms of developing a smart city in Helenski. He found that the city lacked a strategy to reach this goal. He suggested a framework for the same. However, he admitted that there was a priority set by Helenski city in reaching this goal in all the projects initiated.

Krishnan et al. (2020) pointed at the lack of policies as the reason for not realizing the smart cities into digitalized ones. Elberzhager et al. questioned the digitalization of smart cities. They raised a concern on several fronts viz., bringing digitalization based on the needs of users, the solutions pertaining to achieving these needs, bringing the residents into the loop of developing a digitalized smart city, testing the strategy amongst the residents before implementing the same on a large scale.

Adedeji et al. (2022) noted that to bring digitization several challenges must be solved. It is not a one shot investment without any problems. In terms of cost, it is a herculean task to primarily solve all these challenges and then achieve a digitalized smart city. Smart city funds are usually released for such missions without taking the challenges into consideration.

Ylipulli and Luusua (2020) confirmed the need to be a data rich region in order for the implementation of digitalization of smart cities. They credited the public authorities in Helsinki and Espoo for making it data rich before digitalization of these smart cities.

Marcu et al. (2020) opined that digitalization of smart cities was not accelerated since Internet of Things (IoT) was not used in homes and agricultural fields. Rivera et al. (2017) appreciated the use of blockchain to counter cyber fraud. The central reason that perpetrators escape punishment is because of the lack of identity. The authors have stressed the need to introduce blockchain so that data theft and loss privacy can be prevented. This will ultimately lead to establishment of digitalization in smart cities in a safer manner. Wiig (2016) alleged that the words 'smart city' and 'digitalization' were misused. In the names of these words, propaganda was carried out to brand a city that it is 'sophisticated'. Moreover, the residents of the region never got benefited out of it.

Pozdniakova (2017) observed that smart city and digitalization were closely connected with the political and bureaucratic class. It was never a part and parcel of the common man. This kept away people from actively engaging in these activities. Ruohomaa et al. (2019) argued that digitalization and urbanization are not positively correlated. Even smaller cities are capable of achieving digitalization and smart city tag if they intend to.

From the above studies, it is clear that there have been no studies focused on digitalization of a smart city like Bengaluru.

19.3 Discussion

Karnataka received Rs. 5,604 crores from the state and central governments so far, out of which Bengaluru got Rs. 755 crores. This shows that a city bustling with a population of 1,36,08,000 has got a meager slice of the pie (13.48% of the total allocated funds). It shows the lack of political willingness to take the city to the next level. It also indicates that there is a dearth of funds for investing it in the multifarious arms of Smart City Mission.

Bangalore Electricity Supply Company Limited is another entity that is responsible for providing power supply to the city. However, frequent power cuts are a common feature. In a year, a resident of Bengaluru faces on an average 126 h of power outage. An automation project to cut down the power outages in the city led to cost overruns of Rs. 573 crores. The project is still in the pipeline.

The state of Karnataka is set to face election in the month of May 2023. However, political parties are silent about digitalization. The promises and manifestos highlight only the social service schemes.

Akin to power is the next essential source which is broadband connectivity. Bengaluru's broadband connectivity is mainly based on lines hanging from trees or lofty buildings. These lines are affected by rains, which leads to broadband outage. There are no measures taken so far in terms of preventing this. Underground broadband is not safe since water, sewer and sometimes electricity lines run underground. In such a case, there are chances that power or water companies are prone to dig up the underground and cause further outages.

Bureaucratic challenges also persist in terms of digitalization. Bengaluru Metropolitan Transport Corporation is an example. It tried its own to digitalize its ticketing operations. Finally, it was tendered to an outside organization named

'Tummoc' introduced e-ticketing and e-passes. This reduced the work of the conductors. It also brought a sense of relief for passengers who need not board the bus with the exact change to travel to any destination. A year later, Bengaluru Metropolitan Transport Corporation is axing this project. This is how progress in digitalization is stunted in Bengaluru. A daily use digitalization thus came to a halt.

Bangalore Chamber of Industry and Commerce is an organization which can push the government bodies to strive for a higher ranking. It is a body that has acted as a consultant to the government on several occasions. It also has the best interests of many businesses that is operating out of Bengaluru. It can act as a strong chain link between the government and businesses.

A Public Private Partnership is another strategy to bring in more funds into digitalization. This form of strategy is now seen only in the areas of transport infrastructure. The failure of telecom companies prevents companies to invest more into such avenues.

Ride hailing transport companies, E-commerce, Banking and Air/Railway ticketing are the avenues which have seen a steady rise in digitalization due to more Jan Dhan accounts, linking it with Aadhar and mobile phones, demonetization and lack of smaller denomination currencies in the ATM. This has brought in more digitalization in the daily transaction space. Such a digitization happened on a large scale and not specific to a certain area in the country.

A survey conducted by Annual Status of Education Report (ASER) in 2020 at the start of the COVID 19 pandemic confirmed that 68.60% of the household owned smartphones. Owning one does not ensure that people have accepted digitalization. They might use smart phones only for communication purpose.

19.4 Conclusion

Digital transformation needs inputs to transform and does not happen as a magical spell. It needs proper planning and implementation combined with inputs. When there are not enough inputs, the hopes to see the city reach greater ranking in Smart Cities.

The only thing which attracts people to invest in Bengaluru might be the climate. Apart from that, smart city mission has failed to fuel digitalization in the place where it should have blossomed. The condition has gone to such an extent that any upcoming tier 2 or 3 city can easily overtake Bengaluru in the coming years. This will render Bengaluru to be another Singur. Just as Tatas and other companies left West Bengal, Bengaluru may become unfavorable. In order to stop this there must be an intervention from the political and bureaucratic class. It is time to pause and think as to where Bengaluru is headed towards. Mere labeling as Silicon Valley and start up city is not going to help bring in more investments.

This paper does not intend to tarnish the government. It is rather a plea to the government to look into the nuances surrounding the complete digitalization of Bengaluru. While there are avenues which have transformed digitally for good, there are other avenues which are left in the lurch. The implication of the paper majorly falls on the government and the bureaucratic class. Apart from them, it is beneficial to the Non-Government Organizations who strive to bring better governance.

The limitation of this paper is that it has not taken into account any other rankings. It pertains to the 2022 National Smart Cities Mission survey alone. It is not an empirical study. The future scope of research can be done focusing on the specific parts of Bengaluru which house the information technology companies.

References

Adedeji KB, Ponnle AA, Abu-Mahfouz AM, Kurien AM (2022) Towards digitalization of water supply systems for sustainable smart city development—water 4.0. Appl Sci 12(18):9174

Hämäläinen M (2020) A framework for a smart city design: Digital Transformation in the Helsinki Smart City. In: Entrepreneurship and the community: a multidisciplinary perspective on creativity, social challenges, and business, pp 63–86

Krishnan B, Arumugam S, Maddulety K (2020) Critical success factors for the digitalization of smart cities. Int J Technol Manag Sustain Dev 19(1):69–86

Marcu I, Suciu G, Bălăceanu C, Vulpe A, Drăgulinescu AM (2020) Arrowhead technology for digitalization and automation solution: smart cities and smart agriculture. Sensors 20(5):1464

Pozdniakova A (2017) Digitalization process in Ukraine as a prerequisite for the smart city concept development. Baltic J Econ Stud 3(4):206–215

Rivera R, Robledo JG, Larios VM, Avalos JM (2017) How digital identity on blockchain can contribute in a smart city environment. In: 2017 International smart cities conference (ISC2). IEEE, pp 1–4

Ruohomaa H, Salminen V, Kunttu I (2019) Towards a smart city concept in small cities

Vallicelli M (2018) Smart cities and digital workplace culture in the global European context: Amsterdam, London and Paris. City Cult Soc 12:25–34

Wiig A (2016) The empty rhetoric of the smart city: from digital inclusion to economic promotion in Philadelphia. Urban Geogr 37(4):535–553

Ylipulli J, Luusua A (2020) Smart cities with a nordic twist? Public sector digitalization in finnish data-rich cities. Telematics Inform 55:101457

Chapter 20
Investment Perception of Working Age Population Towards Crypto Currency in Sustainable Financial Market

S. Vidhya and T. K. Murugesan

Abstract Crypto currency has gained increasing popularity among the working-age generation as a means of investment and financial transactions. This paper explores the relationship between cryptocurrency and the working age generation, focusing on the attitudes, behaviours, and outcomes associated with digital assets. Using a combination of quantitative and qualitative research methods, we examine the ways in which individuals in the working age range (typically defined as 18–64 years old) interact with cryptocurrency, including their levels of awareness, usage, trading, and mining. This paper examines the relationship between cryptocurrency and the working age generation. It investigates whether members of the working age generation are more likely to use cryptocurrency compared to other age groups and whether they have a better understanding of the technology and its potential uses. The study collected data through a survey of a sample of the working age population and analyzed the results using ANOVA tests. The findings show that while the working age generation has a higher awareness of cryptocurrency compared to older generations, there is no significant difference in their usage or trading behaviour. This suggests that while the working age generation may have a better understanding of cryptocurrency, they are not necessarily more likely to use it compared to other age groups. The study concludes by discussing the implications of these findings for cryptocurrency adoption and future research directions.

Keywords Crypto currency · Working-age generation · Awareness · Usage · Trading · Mining

S. Vidhya (✉) · T. K. Murugesan
Kristu Jayanti College (Autonomous), Bengaluru, India
e-mail: vidhya.s@kristujayanti.com

A. Edward J. et al. (eds.), *Digital Transformation for Business Sustainability*, Contributions to Environmental Sciences & Innovative Business Technology,
https://doi.org/10.1007/978-981-99-7058-2_20

20.1 Introduction

Crypto currency has gained increasing attention in recent years, particularly among younger generations who are more tech-savvy and open to alternative investment options. According to a survey conducted by Gemini, a cryptocurrency exchange, in 2021, 31% of millennials and 22% of Gen Z respondents said they had invested in cryptocurrency, compared to only 9% of Gen X and 2% of baby boomers. One reason for this difference in investment perception may be the familiarity and comfort that younger generations have with technology, including digital currencies. Additionally, the decentralized and non-governmental nature of cryptocurrency aligns with some younger individuals' preference for alternative financial systems outside of traditional institutions. However, crypto currency is still a relatively new and volatile investment option, and it carries significant risks. Cryptocurrency values can fluctuate widely, and there have been instances of high-profile hacks and scams, which can make some investors wary of putting their money into digital currencies. Overall, the investment perception of working age generations towards cryptocurrency may vary based on personal beliefs, experiences, and values, but younger generations appear to be more open to investing in this emerging asset class.

20.2 Review of Literature

According to Voskobojnikov et al. (2020), the massive growth of the cryptocurrency markets in recent years has been accompanied by an influx of new users and investors, bringing the total number of owners into the millions. Moreover, there are already more than 4,900 unique cryptocurrency types. In this growing and chaotic "crypto jungle," new and undiscovered dangers and incentives have an impact on both cryptocurrency users and non-users' behaviour. Prior study has almost exclusively focused on Bitcoin, while mostly ignoring alternative cryptocurrencies and utility tokens.

According to Al-amri et al. (2019), the results demonstrate that scholarly research on the use of cryptocurrencies has grown significantly over the past few years and is still a very active field. The results show that the literature on bitcoin adoption can be divided into three basic categories: qualitative research, quantitative research, and others. The results indicate that little research has been done on the factors that significantly affect whether cryptocurrencies are accepted. Another problem that requires attention is the lack of technical acceptance models. According to Grigoli et al. (2018), there are people of working age in every region of the world. Notwithstanding the recent increase in population ageing, the research claims that many advanced economies have successfully offset their negative impact on labour force participation. In almost half of industrialised economies, the rate of labour force participation increased overall after the global financial crisis.

Kim et al. (2016) used a simple and efficient way to comment on online bitcoin fluctuations in the price of cryptocurrency and the volume of transactions. Three big cryptocurrencies with larger markets and user bases are the focus of the researcher's forecasting. The information in the essay also aided readers in understanding cryptocurrencies from a range of aspects and important views. A wide range of perspectives were interested in the expanding use of crypto currencies.

Economic freedom and the premium on the price of bitcoin are inversely related, according to Viglione. He emphasised the importance of using bitcoin as a reliable source of macroeconomic data and how perfect it was. His methodology raises the possibility of a market-based economic freedom index that is updated frequently rather than once a year and is not determined by experts.

20.3 Statement of the Problem

The increasing popularity of cryptocurrencies has raised questions about how different working age generations view and interact with this new technology. While some studies have explored the relationship between age and cryptocurrency adoption, there is still a need to better understand how different generations approach cryptocurrencies and what factors influence their attitudes and behaviors.

The statement of the problem, therefore, is to investigate how working age generations (i.e., Baby Boomers, Generation X, Millennials Y, and Generation Z) view and use cryptocurrencies. The study aims to identify the factors that influence cryptocurrency adoption among different generations, such as knowledge, risk perception, and trust in the technology. Additionally, the study seeks to explore the potential implications of generational differences in cryptocurrency adoption for the financial industry and society.

Understanding how different working age generations view and interact with cryptocurrencies is crucial for policymakers, financial institutions, and cryptocurrency advocates, as it can inform strategies for promoting wider adoption and addressing potential barriers to adoption. The study will provide valuable insights into the attitudes and behaviours of different generations towards cryptocurrencies and contribute to a better understanding of the role of cryptocurrencies in the financial ecosystem.

The common notion of risk and returns lies on the tripod of investment i.e.; land, gold and stocks. There is an established choice of investment for each generation. However, the researches that are already undertaken does not provide any insight to the perception of such generations to the latest addition to the investment tripod called crypto currency.

20.4 Objective of the Study

To understand the differences in the investment perception of different working age generations towards cryptocurrency.

20.5 Sampling

Sampling is the process of selecting a representative group of the overall population to be considered for the study in order to arrive at a conclusion on the full population. When there is a huge population, sampling is conducted. Since there is a relatively huge population for the research and sampling methods are used only 106 people were included in the study's sample.

20.6 Data Analysis and Interpretation

The primary data has been analysed with the help of statistical tools like correlation and regression. The variables taken for the study is Age Group, Awareness Level, Time period of awareness, usage, trading and mining; with age group as the independent variables and the other variables as the dependent variables (Table 20.1).

Inference: Based on the correlation table provided, there are several significant correlations between the variables. Firstly, there is a significant positive correlation between Age Group and Awareness Level ($r = 0.199$, $p = 0.041$), indicating that as age group increases, so does the awareness level. Secondly, there is a significant negative correlation between Awareness Level and Usage ($r = -0.575$, $p = 0.000$), Trading ($r = -0.550$, $p = 0.000$), and Mining ($r = -0.470$, $p = 0.000$). This suggests that as the level of awareness increases, the usage, trading, and mining of the product or service decreases. Thirdly, there is a significant positive correlation between Usage and Trading ($r = 0.903$, $p = 0.000$) and between Usage and Mining ($r = 0.651$, $p = 0.000$). This indicates that there is a strong positive relationship between usage and trading, as well as usage and mining. Lastly, there is a significant positive correlation between Trading and Mining ($r = 0.721$, $p = 0.000$), indicating that as trading increases, so does mining. Overall, these correlations suggest that age group and awareness level are important factors in determining usage, trading, and mining of the product or service. Additionally, usage is strongly related to trading and mining, and trading is related to mining. These findings could have implications for marketing and advertising strategies aimed at different age groups and increasing awareness levels (Tables 20.2 and 20.3).

Inference: The ANOVA table and coefficients table provided show the results of a regression analysis conducted to examine the relationship between Age Group and

Table 20.1 Correlation among variables

Correlations

		Age group	Awareness level	Time period of awareness	Usage	Trading	Mining
Age group	Pearson correlation	1	0.199*	0.011	−0.024	0.042	0.053
	Sig. (2-tailed)		0.041	0.908	0.808	0.670	0.589
	N	106	106	106	106	106	106
Awareness level	Pearson correlation	0.199*	1	−0.348**	−0.575**	−0.550**	−0.470**
	Sig. (2-tailed)	0.041		0.000	0.000	0.000	0.000
	N	106	106	106	106	106	106
Time period of awareness	Pearson correlation	0.011	−0.348**	1	0.267**	0.187	0.174
	Sig. (2-tailed)	0.908	0.000		0.006	0.055	0.075
	N	106	106	106	106	106	106
Usage	Pearson correlation	−0.024	−0.575**	0.267**	1	0.903**	0.651**
	Sig. (2-tailed)	0.808	0.000	0.006		0.000	0.000
	N	106	106	106	106	106	106
Trading	Pearson correlation	0.042	−0.550**	0.187	0.903**	1	0.721**
	Sig. (2-tailed)	0.670	0.000	0.055	0.000		0.000
	N	106	106	106	106	106	106
Mining	Pearson correlation	0.053	−0.470**	0.174	0.651**	0.721**	1
	Sig. (2-tailed)	0.589	0.000	0.075	0.000	0.000	
	N	106	106	106	106	106	106

* Correlation is significant at the 0.05 level (2-tailed)
** Correlation is significant at the 0.01 level (2-tailed)

Awareness Level. The ANOVA table shows that the regression model is significant ($F = 4.279$, $p = 0.041$), indicating that Age Group significantly predicts Awareness Level. The model accounts for 2.243 units of the variance in Awareness Level, with the remaining 54.521 units of variance unexplained by the model.

The coefficients table shows the unstandardized coefficients for the constant and Age Group, as well as the standardized coefficient (beta) for Age Group. The constant

Table 20.2 ANOVA table among age group and awareness level

Model		Sum of squares	df	Mean square	F	Sig.
1	Regression	2.243	1	2.243	4.279	0.041[b]
	Residual	54.521	104	0.524		
	Total	56.764	105			

[a] Dependent Variable: Awareness Level
[b] Predictors: (Constant), Age Group

Table 20.3 Coefficient table among age group and awareness level

Model		Unstandardized coefficients		Standardized coefficients	t	Sig.
		B	Std. error	Beta		
1	(Constant)	1.576	0.195		8.067	0.000
	Age group	0.179	0.087	0.199	2.069	0.041

[a] Dependent Variable: Awareness Level

(1.576) represents the predicted value of Awareness Level when Age Group is zero. The coefficient for Age Group (0.179) indicates that for every one-unit increase in Age Group, Awareness Level is predicted to increase by 0.179 units. The standardized coefficient (beta) for Age Group (0.199) indicates that Age Group has a small to moderate effect on Awareness Level.

In summary, the regression analysis suggests that Age Group is a significant predictor of Awareness Level, with older age groups being associated with higher levels of awareness. However, the model only accounts for a small amount of the variance in Awareness Level, and other factors may also contribute to awareness levels (Tables 20.4 and 20.5).

Inference: The ANOVA table and coefficients table provided show the results of a regression analysis conducted to examine the relationship between Age Group and Usage.

The ANOVA table shows that the regression model is not significant ($F = 0.059$, $p = 0.808$), indicating that Age Group does not significantly predict Usage. The model

Table 20.4 ANOVA table among age group and usage

Model		Sum of squares	df	Mean square	F	Sig.
1	Regression	0.012	1	0.012	0.059	0.808[b]
	Residual	20.592	104	0.198		
	Total	20.604	105			

[a] Dependent Variable: Usage
[b] Predictors: (Constant), Age Group

Table 20.5 Coefficient table among age group and usage

Model		Unstandardized coefficients		Standardized coefficients	t	Sig.
		B	Std. error	Beta		
1	(Constant)	1.291	0.120		10.757	0.000
	Age group	−0.013	0.053	−0.024	−0.244	0.808

[a] Dependent Variable: Usage

accounts for only 0.012 units of the variance in Usage, with the remaining 20.592 units of variance unexplained by the model.

The coefficients table shows the unstandardized coefficients for the constant and Age Group, as well as the standardized coefficient (beta) for Age Group. The constant (1.291) represents the predicted value of Usage when Age Group is zero. The coefficient for Age Group (−0.013) indicates that for every one-unit increase in Age Group, Usage is predicted to decrease by 0.013 units. However, since the model is not significant, this coefficient is not meaningful.

In summary, the regression analysis suggests that Age Group is not a significant predictor of Usage, and the model does not explain much of the variance in Usage. Other factors may be more important in predicting Usage, and further research is needed to determine these factors (Tables 20.6 and 20.7).

Inference: In Table, the ANOVA table shows the results of the analysis of variance for the relationship between Age Group and Trading. The regression model's sum of squares is 0.033, with 1 degree of freedom and a mean square of 0.033. The F-statistic is 0.183, and the associated p-value is 0.670, indicating that there is no significant

Table 20.6 ANOVA table among age group and trading

Model		Sum of squares	df	Mean square	F	Sig.
1	Regression	0.033	1	0.033	0.183	0.670[b]
	Residual	18.533	104	0.178		
	Total	18.566	105			

[a] Dependent Variable: Trading
[b] Predictors: (Constant), Age Group

Table 20.7 Coefficients table among age group and trading

Model		Unstandardized coefficients		Standardized coefficients	t	Sig.
		B	Std. Error	Beta		
1	(Constant)	1.181	0.114		10.369	0.000
	Age group	0.022	0.051	0.042	0.428	0.670

[a] Dependent Variable: Trading

relationship between Age Group and Trading. The above displays the coefficients for the regression model predicting Trading from Age Group. The intercept term's coefficient is 1.181, indicating that the expected value of Trading when Age Group is 0 is 1.181. The coefficient for Age Group is 0.022, suggesting that for each one-unit increase in Age Group, the expected value of Trading increases by 0.022. However, the t-statistic for Age Group is 0.428, and its associated p-value is 0.670, indicating that this relationship is not statistically significant. Therefore, there is insufficient evidence to conclude that Age Group is a significant predictor of Trading (Tables 20.8 and 20.9).

Inference: In the Table, we see the ANOVA table for the relationship between age group and mining. The regression model has a sum of squares of 0.034 and a mean square of 0.034 with 1 degree of freedom. The F-ratio is 0.294 with a p-value of 0.589, indicating that there is no significant relationship between age group and mining. We see the coefficients table for the same relationship. The intercept has an unstandardized coefficient of 1.085 and a standard error of 0.092. The coefficient for age group is 0.022 with a standard error of 0.041 and a beta value of 0.053. The t-value is 0.542 with a p-value of 0.589, indicating that there is no significant relationship between age group and mining.

Table 20.8 ANOVA table among age group and mining

Model		Sum of squares	df	Mean square	F	Sig.
1	Regression	0.034	1	0.034	0.294	0.589[b]
	Residual	12.117	104	0.117		
	Total	12.151	105			

[a] Dependent Variable: Mining.
[b] Predictors: (Constant), Age Group

Table 20.9 ANOVA table among age group and mining

Model		Unstandardized coefficients		Standardized coefficients	t	Sig.
		B	Std. Error	Beta		
1	(Constant)	1.085	0.092		11.787	0.000
	Age group	0.022	0.041	0.053	0.542	0.589

[a] Dependent Variable: Mining.

20.7 Conclusion

Crypto currencies have become increasingly popular over the past decade, particularly among younger generations. Here is the inference of how different working age generations view and interact with crypto currencies: Generation X (born 1965–1980): Generation Xers are more likely to have some exposure to cryptocurrencies, but are still cautious in their approach. They may be more likely to invest in established cryptocurrencies like Bitcoin, but may be wary of the volatility and lack of regulation in the market. Millennials (born 1981–1996): Millennials are the generation most likely to embrace cryptocurrencies. They have grown up with technology and are comfortable using digital currencies for transactions. Many Millennials view cryptocurrencies as an alternative to traditional banking and see the potential for them to disrupt the financial industry. Generation Z (born 1997–2012): Generation Z is the youngest working age generation, but they are already showing a keen interest in cryptocurrencies. Like Millennials, they are digital natives and are comfortable using digital currencies. They may also view cryptocurrencies to make quick profits. Overall, the younger generations are more likely to be open to cryptocurrencies and see the potential for them to transform the financial industry. However, all generations should be cautious when investing in cryptocurrencies due to the volatility and lack of regulation in the market. The reports regarding crypto currency and allied transactions were highly confidential due to government laws and regulations. There are constraints with regards to time and cost for the analysis. The scope of the study is confined to identify the style and pattern of investing of the three generations under study in Bengaluru, Karnataka. This research also aims to identify the differences in the behaviour towards crypto currency, that is to find out to what extent the generation gap affects an individual's financial decisions and to get a brief picture about the general risk attitude of the three generations. This research helps to identify the various sources where individuals get to invest in bitcoin and what factors influence them to take decisions regarding investment.

References

Al-amri NH, Zakaria AH, Hassan S (2019) Cryptocurrency adoption: current stage, opportunities, and open challenges. Int J Adv Comp Res 9(44):293–307

Grigoli F, Koczan Z, Topalova P (2018) A cohort-based analysis of labor force participation for advanced economies. IMF working paper

Kim YB, Kim JG, Kim W, Im JH, Kim TH, Kang SJ, Kim CH (2016) Predicting fluctuations in cryptocurrency transactions based on user comments and replies

Voskobojnikov A, Obada-Obieh B, Huang Y, Beznosov K (2020) Surviving the cryptojungle: perception and management of risk among North American cryptocurrency. In: International conference on financial cryptography and data security

SPRINGER NATURE

GPSR Compliance

The European Union's (EU) General Product Safety Regulation (GPSR) is a set of rules that requires consumer products to be safe and our obligations to ensure this.

If you have any concerns about our products, you can contact us on ProductSafety@springernature.com

In case Publisher is established outside the EU, the EU authorized representative is:

Springer Nature Customer Service Center GmbH
Europaplatz 3
69115 Heidelberg, Germany

The manufacturer's authorised representative in the EU is Springer Nature Customer Service Centre GmbH, Europaplatz 3, 69115 Heidelberg, Germany. If you have any concerns regarding our products, please contact ProductSafety@springernature.com

Printed and bound by CPI Group (UK) Ltd, Croydon, CR0 4YY
30/09/2026
02225322-0001